Physical Security Professional (PSP) Reference

Excerpts from the *Protection of Assets (POA) Manual*

(SECOND EDITION)

Copyright © 2009 by ASIS International

ISBN 978-1-887056-74-8

All rights reserved. No part of this publication may be reproduced, stored in a retrieval system, or transmitted, in any form or by any means, electronic, mechanical, photocopying, recording, or otherwise, without the prior written consent of the copyright owner.

Printed in the United States of America

10 9 8 7

TABLE OF CONTENTS

PART 1 – INTRODUCTION TO ASSETS PROTECTION

1.1. Introduction 1

1.2. Basis for Enterprise Assets Protection 2
    1.2.1 Defining Assets Protection 2
    1.2.2 Relation to Security and Other Disciplines 3
    1.2.3 Historical Perspectives 4

1.3. Current Practice of Assets Protection 7
    1.3.1 Underlying Principles 8
    1.3.2 Assets Protection in Various Settings 9

1.4. Forces Shaping Assets Protection 14
    1.4.1 Technology and Touch 14
    1.4.2 Globalization in Business 16
    1.4.3 Standards and Regulation 17
    1.4.4 Convergence of Security Solutions 20
    1.4.5 Homeland Security and the International Security Environment 21

1.5. Management of Assets Protection 21
    1.5.1 Concepts in Organizational Management 22
    1.5.2 Management Applications in Assets Protection 24
    1.5.3 Security Organization within the Enterprise 25

1.6. Behavioral Issues in Assets Protection 26
    1.6.1 Behavioral Science Theories in Management 27
    1.6.2 Applications of Behavioral Studies in Assets Protection 28

Appendix A 31

References 44

PART 2 – CLOSED-CIRCUIT TELEVISION SYSTEMS

2.1. Introduction 47
2.2. Theory of Visual Security 48
2.3. Reasons for CCTV in Security 48
2.4. Analog System Components 51
2.5. Digital System Components 53
2.6. System Design 54
2.7. Equipment Selection 62
2.8. Camera Formats and Lenses 66
2.9. Switching Systems 68
2.10. Recording Systems 71
2.11. Where CCTV is Heading 73

PART 3 – HIGH-RISE STRUCTURES

- 3.1. Introduction 75
  - 3.1.1. What is High-Rise Structure? 75
- 3.2. Life Safety Considerations.. 76
  - 3.2.1. Special Concerns of High-Rise Structures 76
  - 3.2.2. Dealing with the Life Safety Problem 80
- 3.3. Security Considerations 87
  - 3.3.1. Special Concerns of High-Rise Structures 87
  - 3.3.2. Life Safety and Security Dilemma 88
  - 3.3.3. Typical High-Rise.. 88
  - 3.3.4. Building Operating Modes 90
  - 3.3.5. Building Elements.. 90
  - 3.3.6. Access Control of Public Areas 91
  - 3.3.7. Access Control of Interior Floors and Spaces.. 96
  - 3.3.8. Access Control of Building Maintenance Spaces.................................. 97
  - 3.3.9. Access Control of Air Intakes and Telecommunication Services 98
  - 3.3.10. Security Features 99
- 3.4. Summary 109
- References 110

Part 4 – INTEGRATED SECURITY SYSTEMS DESIGN AND SPECIFICATION

- 4.1. Introduction 113
- 4.2. Systems Design Process 115
- 4.3. Planning and Assessment Phase 116
  - 4.3.1 Requirements Analysis... 118
  - 4.3.2 Design Requirements 121
  - 4.3.3 Basis of Design 123
  - 4.3.4 Conceptual Design 123
  - 4.3.5 Design Criteria 125
  - 4.3.6 Design Team 131
- 4.4. Design and Documentation Phase.. 132
  - 4.4.1 Contractual Details 133
  - 4.4.2 Specifications 133
  - 4.4.3 Drawings... 135
  - 4.4.4 Design Coordination 139
  - 4.4.5 Construction Document Review, Approvals, and Issue 142
  - 4.4.6 Procurement Phase 143
  - 4.4.7 Sole Source Procurement.. 144
  - 4.4.8 Request for Proposal... 144
  - 4.4.9 Invitation for Bid 145
  - 4.4.10 Procurement Process 145
- References 148

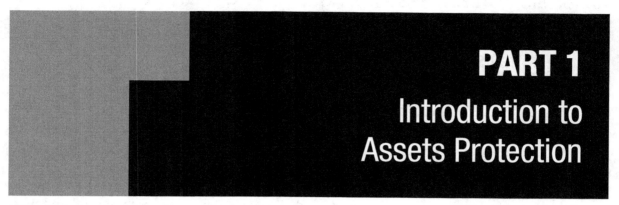

This text is excerpted from the March 2008 revision, published as "Part I: Introduction to Assets Protection" in the "Security Management" chapter of the *Protection of Assets (POA) Manual*.

1.1. INTRODUCTION

Protecting an organization's assets is a daunting task. The business world, the security arena, and life itself are changing at lightning speed. Globalization, information technology, instant communications, complex and asymmetric threats, public opinion, mergers and acquisitions, conglomerates and partnerships, and regulation all have a major influence on how security professionals must perform their mission. In addition to needing a broad array of security expertise, today's security professional must be an adaptable, strategic thinker, skilled in process management and fast, accurate program implementation.

The *Protection of Assets Manual* is designed as a support tool for security professionals and others with similar responsibilities. It provides information on all aspects of security and related functions and helps readers balance costs and results in planning, developing, and implementing sound risk management strategies.

Because of the rapid pace of change, the *POA Manual* is a living document. It features periodic updates and guides readers to other sources for further information.

1.2. BASIS FOR ENTERPRISE ASSETS PROTECTION

1.2.1 DEFINING ASSETS PROTECTION

For many people, the term *assets protection* suggests finance. Security professionals, however, think of assets protection in a different, broader sense. In the security arena, one often speaks of protecting three types of assets: people, property, and information. The larger view of assets protection, however, also considers intangible assets, such as an organization's reputation, relationships, and creditworthiness.

In considering all of an organization's assets and all potential hazards, both natural and man-made, the security function should take the lead on some matters and play a supporting role in others. This approach helps ensure that the security function is, and is seen to be, a value-adding element of the organization. The greatest protection of corporate assets occurs when the appropriate mix of physical, procedural, and electronic security measures are in place equal to the value of the assets being protected. This creates an effective defense in-depth asset protection program.

Graduate students in a security management program were recently asked to define assets protection from their perspective. The students were all experienced, mid-career professionals in security, law enforcement, or the military. Almost all the students mentioned elements as asset definition, threat assessment, vulnerability and risk analysis in deterring security methods to reduce risk mitigation. Plus, the need to balance security costs with the benefits of protective measures employed. However, several additional aspects of assets protection emerged as well:

- Both tangible and intangible assets must be considered.
- A key objective is maintaining smooth business operations.
- Post-incident business or mission continuity is an important element.
- Both the current and future risk environments must be considered.
- Providing a safe and healthy environment should be factored in.
- Liability reduction/management is an important component.

As those students seemed to understand, assets protection must be a comprehensive, proactive function that is directly tied to the organization's mission.

In addition, it is essential to know what needs to be protected. In many cases, asset owners (such as business owners or managers) lack a thorough understanding of what their real assets are. Some think purely in financial terms, while others focus on tangible goods, such as facilities, inventory, vehicles, or equipment. A wider view of assets might include those listed in Figure 1.

1.2.2 RELATION TO SECURITY AND OTHER DISCIPLINES

Because assets protection is a broad, complex function, many departments or elements of an organization may be involved in it. However, a single office or person should be designated as the assets protection focal point. Assets protection professionals should either lead or follow, but in either case they should not allow themselves to be left out of key deliberations and decisions. Though it is the responsibility of senior management to provide the resources needed to enhance the protection of assets, it is the asset protection professional's responsibility to provide them with the best information for their decision-making process.

FIGURE 1
Examples of Organizational Assets by Type

TANGIBLE	INTANGIBLE	MIXED
Facilities/buildings	Reputation/image	People
Equipment	Goodwill/trust	Intellectual property
Inventory	Brand recognition	Knowledge
Vehicles	Relationships	Proprietary processes
Raw materials	Vendor diversity	Information technology
Cash/money	Longevity/history	capabilities
Accounts receivable	Past performance	Land/real estate
Supplies/consumables	Experience	Infrastructure
Telecommunications systems	Quality assurance processes	Credit rating/financial stability
Other capital assets	Workforce morale/spirit/loyalty	Customers (customer base)
	Workforce retention	Contracts in place
	Management style	Financial investments
	Human capital development	Geographic location
	Liaison agreements	Staffing sources/recruiting
	Market share	Certifications (e.g., ISO 9000)
		Continuity posture/resiliency
		Safety posture

NOTE: Tangible assets are generally those one can see, touch, or directly measure in physical form. Mixed assets have both tangible and intangible characteristics.

Assets protection incorporates all security functions as well as many related functions, such as investigations, risk management, safety, quality/product assurance, compliance, and emergency management. Therefore, the senior assets protection professional must have strong collaboration and coordination skills as well as a thorough understanding of the workings of the enterprise. In today's asset protection program, countermeasures need to include people, hardware, and software.

Of particular interest today is convergence, which is the "integration of traditional and information [systems] security functions" (ASIS International, 2005). Such convergence makes collaboration even more important.

1.2.3 HISTORICAL PERSPECTIVES

From the dawn of mankind, organizations have faced threats to their safety and security. One of the tribe's important functions was the protection of its assets, which might include land, crops, water supplies, or its cultural or religious heritage.

Over the centuries, upon arriving in a new country, immigrants from particular regions have tended to settle together in communities that became known as ghettos. These ghettos have had a strong assets protection aspect.

Like tribes, gangs today emphasize assets protection. Their assets may include "turf," recognition, members, weapons, or market share of illegal activities.

Families, too, protect their assets, which include family members, the home and its contents, vehicles, financial assets, pets, occupations, and status in the community. Families use such methods as security equipment, insurance, education, communications procedures, and neighborhood watch groups.

Different assets protection methods work in different situations (Webster University, 2006):

> The protection of assets is not an exact science. What works in one situation may have disastrous results in another. Asset owners and security professionals alike must analyze specific situations or environments; recognize needs, issues and resources; and draw conclusions regarding the most appropriate protection strategies and applications.

Assets protection can be performed by internal entities, external entities, or a combination. In the United States, the first private security firms emerged in the mid-19th century. They began as investigative agencies and expanded to provide other assets protection functions, such as executive protection, intelligence collection, counterintelligence, cargo escort, and protection of railroads, a critical infrastructure of the day (Securitas, 2006).

The concepts, techniques, tools, and philosophies of assets protection change as threats mutate, technologies advance, management approaches develop, and business around the world becomes transformed.

Influences in Assets Protection

Many recent developments have affected the practice of assets protection. In the early 1970s, for example, computer security began to flourish as a separate discipline (National Institute of Standards and Technology, 2006) because of society's increasing reliance on information systems.

Another influence was the recognition of the vulnerability of critical infrastructure to both natural and intentional attacks. In the United States, critical infrastructure was initially defined as comprising the following industry sectors: transportation, oil and gas, water, emergency services, government services, banking and finance, electrical power, and telecommunications. More sectors were added later. Significantly, most U.S. critical infrastructure is owned or operated by private enterprises. In the United States, attention to the security of critical infrastructure increased greatly after the 1993 attack on the World Trade Center in New York City and the bombing of the Alfred P. Murrah Federal Building in Oklahoma City two years later.

Damage to the Pentagon caused by the September 11th attack.

Photograph by Kevin Peterson

To security professionals, the terrorist attacks of September 11, 2001, represented the most significant turning point in assets protection around the world. That attack

- led to increased security budgets and reduced constraints on security policies and procedures
- fostered communication between security officials and front-office executives
- enhanced threat awareness and vigilance by business managers and employees

In some cases, knee-jerk reactions to 9/11 wasted valuable resources. For example, one company with facilities in several countries ordered each site to post a security officer at its entrance. However, the new security officers had no idea of their roles and responsibilities and had no way to communicate with other security staff at the sites. At best they were able to provide a false sense of security. Similarly, after 9/11 many organizations spent much more than necessary on security technology.

The shock of 9/11 also caused an overemphasis—in terms of security solutions—on terrorist attacks instead of the broader spectrum of realistic security risks. Even now, resources that could have been dedicated to information technology (IT) security, information asset protection, and traditional crime or loss prevention is being diverted to antiterrorism measures, such as blast-resistant materials, stand-off zones, bollards, chemical/biological hazard sensors, and similar items. Even in school security, interest in traditional, comprehensive assets protection has often given way to preparation for terrorist attacks.

Over time, the 9/11 attacks have partly redefined assets protection. The following are some of the beneficial changes:

- a change in public expectations and an increase in the level of security measures that the public will tolerate
- an ongoing examination of personal privacy versus public protection
- more serious study of security and protective services budgets and strategies
- better information sharing within and between the security and law enforcement communities have improved crime-fighting capabilities
- greater application of advanced technologies to threat analysis, vulnerability assessment, information sharing, and protective measures
- more widespread discussion of strategic protection concepts incorporating risk management and comprehensive assets protection
- more emphasis on security and assets protection research

Similarly, the 2001 anthrax scare in the United States led to much greater emphasis on the security of mailroom operations. In addition, the Sarbanes-Oxley Act in the United States has required publicly traded corporations to perform more extensive assessment and reporting.

INTRODUCTION TO ASSETS PROTECTION

Respondents to one security-related survey rated the act as the second most important legislation having a moderate or major impact on their organization (ASIS International, 2005, p. 48).

Patterns of Change

In assets protection, the period between major paradigm shifts (including technological developments and conceptual shifts) has been decreasing. As Figure 2 shows, during the 1950s and 1960s several years passed between major paradigm shifts. In more recent decades, the interval between those shifts has decreased to the point where changes today follow each other rapidly.

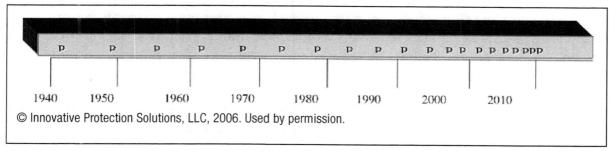

FIGURE 2
Paradigm Shift Frequency Model

© Innovative Protection Solutions, LLC, 2006. Used by permission.

These paradigm shifts include changes in surveillance technology, integrated security systems, the scope of security professionals' duties, legal and liability issues, the regulatory environment, the use of computers in the security function, public/private partnerships, antiterrorism, convergence, and global business relationships. Security professionals must be prepared for rapid change in the workplace.

Another change is that assets protection is increasingly based on the principle of risk management, a term rather recently applied to security management and assets protection (Webster University, 2006). The ASIS International's 2006 *General Risk Security Guideline* define "risk" as the possibility of loss resulting from a threat, security incident, or event. The concept is a perfect fit for assets protection, the primary objective of which is to manage risks by balancing the costs and benefits of protection measures.

1.3. CURRENT PRACTICE OF ASSETS PROTECTION

This section discusses two important issues in assets protection: the field's underlying principles and the practice of assets protection in various industry sectors.

1.3.1 UNDERLYING PRINCIPLES

One framework for viewing the underlying principles of assets protection states that three concepts form a foundation for any assets protection strategy. Those concepts are known as the five avenues to address risk, balancing security and legal considerations, and the five Ds.

Five Avenues to Address Risk

This concept contends that there are five distinct avenues for addressing identified risks to assets: risk avoidance, risk transfer, risk spreading, risk reduction, and risk acceptance. Carefully considering these avenues is an effective way for assets protection professionals and management to think creatively in designing ways to protecting assets.

Balancing Security and Legal Considerations

Organizations need to find the right balance between a security approach and a "legal" approach. Some enterprises rely entirely on legal measures, such as patents, copyrights, trademarks, and service marks, to protect their critical information. They mistakenly believe that with these legal protections in place, they do not need stringent security programs. Alternatively, some executives believe a strong security program eliminates the need for legal measures. Of course, both types of measures are needed. The legal approach must also consider when and how incidents will be litigated, what preliminary measures must be in place for successful litigation, and how litigation costs will be managed.

The Five Ds

This security approach complements the "legal" approaches discussed above. In this concept, the first objective in protecting assets is to *deter* any type of attack. The second objective is to *deny* the adversary access to the asset, typically through traditional security measures. The third objective, if the first two fail, is to *detect* the attack or situation, often using surveillance and intrusion detection systems, human observation, or a management system that identifies shortages or inconsistencies. Once an attack or attempt is in progress, the fourth objective is to *delay* the perpetrator through the use of physical security and target hardening methods, or use of force. Finally, in today's terrorist environment with more violent criminals, it may become necessary to *destroy* the aggressor if the situation warrants it.

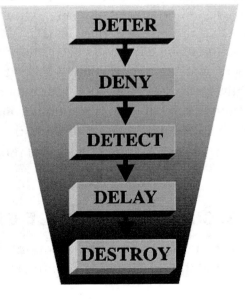

In short, assets protection should involve a comprehensive strategy, not just piecemeal elements (officers, closed-circuit television, access control systems, etc.).

INTRODUCTION TO ASSETS PROTECTION

1.3.2 ASSETS PROTECTION IN VARIOUS SETTINGS

Many security principles and procedures are common across sectors, geographic areas, and various sizes and types of organizations. However, each particular industry has its own culture, environment, and issues that influence assets protection.

Health Care Sector

Hospitals are open to the public 24/7 and tend to have an open environment. Patients are vulnerable, and hospitals can be a high-stress environment for all concerned: patients, visitors, and staff.

Hospitals also have to be concerned about information assets, especially patient privacy, the protection of which is often governed by regulation, such as, in the United States, the Health Insurance Portability and Accountability Act (HIPAA) and criteria set by the Joint Commission on Accreditation of Healthcare Organizations (JCAHO). In addition, many health care institutions, especially at universities, engage in medical research, an activity that calls for protection of sensitive information, intellectual property, facilities, and materials. Assets protection staff may also need to focus on maintaining the hospital's reputation, another key asset.

The most serious threats in health care involve workplace and domestic violence, threats, harassment, internal theft, vandalism, extremist activity, fraud, threats to high-risk or high-profile patients, and violence in emergency departments.

Health care security professionals can gain management support through these means (Stewart, 2006):

- demonstrating a knowledge of hospital management issues and respecting the business aspects of the enterprise
- maintaining a dialogue with management to ensure they understand the hospital's risks and vulnerabilities, as well as the assets protection program itself

Whether security officers in health care settings should be armed is the subject of ongoing debate.

Educational Sector

Educational institutions range from preschools to universities and include both public and private institutions. Schools at all levels have historically been viewed as somewhat insulated from the ills of society, but in recent years more attention has been paid to school security.

At the lower academic levels, security responsibility may fall under the school board, county or city, or local police department. Most colleges and universities maintain their own security function, which may or may not be connected to the campus police department.

Educational institutions face a wide range of threats, such as assaults against students and staff, facility damage, vandalism, theft of goods (computers, equipment, supplies, etc.), theft of private information, attacks against IT, white-collar crime, liability, and natural disasters. Universities also face the theft of research information.

At most schools, much of a security director's time is spent on crisis management. Evacuation planning, preparations for shelter-in-place situations, liaison with first responders, awareness, training, and exercises are all critical in that environment. In addition, schools may be called on to serve as community shelters or medical triage centers during disasters. Figure 3 lists some of the common security issues at each educational level.

Universities include more than classrooms—they may also feature dormitories, restaurants, stores, libraries, entertainment venues (clubs, theaters, bowling alleys, fitness centers, game rooms, etc.), sporting facilities, worship centers, conference centers, and hospitals. Further security issues are raised by the fact that some students may be living away from home for the first time and may not behave as well as they should or show the right level of safety and security consciousness. Universities also host many students from other countries, who may violate bans on certain exports or may overstay their visas.

High crime rates, high-profile incidents, and a questionable campus safety record can harm a university's image and lead to a loss of students, revenue, grant money, and research projects.

Security directors in the educational environment must take a comprehensive risk management approach to their assets protection program. In their security planning, they should consider many factors, such as the size and demographics of the school, the characteristics of the surrounding area, the mission and culture of the institution, the types and values of assets, the school's image, its management style, and any identifiable threats.

Fast Food Sector

This sector, also known as the quick-service restaurant (QSR) industry, features many company-owned restaurants and franchise stores around the world. The largest companies often have an in-country or regional assets protection director, who reports to the local business unit head and the corporate assets protection director. The wide geographical dispersion also makes QSRs vulnerable to varying levels of ordinary crime, activism, vandalism, and terrorism. Companies in this industry work hard to protect the value of their brand.

FIGURE 3
School Security Considerations

Level	Considerations
Preschool	Health and safety Teacher/staff backgrounds Constant student oversight Potential for parental/stranger abduction
Elementary (K through 8)	Student oversight Teacher/staff backgrounds Inappropriate discipline Early gang and drug abuse prevention Exposure to inappropriate issues Student interrelationships
Secondary and High School	Student independence/student interrelationships Teacher/staff backgrounds Teacher/staff relationships with students Gang and drug/alcohol abuse prevention Exposure to inappropriate issues Weapons and contraband exclusion Facility access control Protection of equipment, chemicals, other resources
College and University	Students as an asset and a threat Lifestyle (student independence, drugs, alcohol, etc.) Residential setting Multiple facilities (retail, food service, entertainment) Overall crime environment Potential for hate crimes and activist groups Sports and entertainment venues Laboratory/research facilities and information

© Innovative Protection Solutions, LLC, 2006. Used by permission.

The industry emphasizes cost control, margins, and profit and loss management. Thus, assets protection professionals must focus on theft prevention, anti-fraud programs, strategic planning, and supply chain/vendor/distribution integrity. The QSR industry employs a range of security technology, including closed-circuit television (CCTV) tied to point-of-sale systems (e.g., cash registers). Assets protection teams in the industry also investigate suspected false claims of employee or customer injuries.

Because of the high employee turnover rate and the geographic dispersion of stores, security training is both essential and difficult. Modern IT can enhance the company's ability to conduct safety and security training—for example, by facilitating distance learning. One focus of employee training is simply teaching whom to call and how to report suspicious activity. Most companies maintain toll-free hot lines. In addition, employee awareness can be bolstered using security posters, changed regularly.

Telecommunications Sector

Assets protection in the telecommunications sector has changed in the wake of industry deregulation; the boom in wireless, Internet, fiber optic, and other telecommunications technologies; and, in the United States, the designation of the telecommunications system as a national critical infrastructure. Assets protection in the telecom sector now encompasses four major areas:

- **Information security:** protecting competitive and proprietary information; protecting information about the telecommunication infrastructure; and protecting voice and data signals
- **Network and computer security:** protecting networks from hacking and other forms of cyber attacks; protecting computers and other equipment from viruses
- **Fraud prevention:** protecting the company from toll fraud, calling card misuse, and other frauds
- **Physical security:** protecting the people, places, and things that make telecommunications networks function

Assets protection in telecommunications is greatly affected by government regulation. Some jurisdictions mandate specific security practices, limiting the ability of assets protection managers to tailor programs to their particular environment. Another security challenge arises from the wide exposure of the industry's product (electronic signals), which are susceptible to both physical and electronic threats. Finally, telecom companies' fiber and cables are often routed through or under property owned by others. Therefore, assets protection strategies must consider property rights and access issues.

Aerospace Sector

The aerospace sector, which includes civil aircraft, military aircraft, missiles, space systems, and aerospace services, is characterized by fierce, global competition; large, complex contracts; international joint ventures; and a huge network of vendors, all of which factors significantly complicate assets protection strategies.

INTRODUCTION TO ASSETS PROTECTION

NASA Photo

In addition to traditional corporate safeguards, firms in this sector should consider the following:

- protection of sensitive, proprietary, and export-controlled technical information
- handling of government classified information
- regulatory and reporting compliance at the local, national, and international levels
- integration of safety and security programs
- domestic and international travel security
- test and evaluation program security

The larger aerospace firms maintain large security departments staffed with various security specialties. By contrast, small aerospace vendors often have no security resources. Therefore, it is best to discuss security support at the outset of a new project and agree who will be responsible for various aspects of assets protection and what resources each player will contribute.

Assets protection in the aerospace industry is also affected by the climate of risk taking; the extent of high-value information that must be protected; and the industry's high profile, which attracts adversaries in the form of competitors, activist groups, and white-collar criminals.

These industry snapshots illustrate the wide variety of issues, concerns, and environmental factors that affect assets protection programs. They highlight the meshing of security concerns with business and management issues in planning for a safe and secure setting in which to conduct the enterprise's mission.

1.4. FORCES SHAPING ASSETS PROTECTION

This section examines five forces that are shaping the practice of assets protection:

- technology and touch
- globalization in business
- standards and regulation
- convergence of security solutions
- homeland security and the international security environment

Some of these forces are at least partially within an assets protection manager's ability to influence, while others are not. In either case, security professionals should study and leverage these forces as they formulate tomorrow's protective strategies.

1.4.1 TECHNOLOGY AND TOUCH

Assets protection has always required a balance between human and technological solutions. Sometimes the balance swings too far toward technology. The following statements are described as symptoms of "high-tech intoxication" (Naisbitt, 1999):

- We look for the quick fix.
- We fear and worship technology.
- We blur the distinction between real and fake.
- We accept violence as normal.
- We love technology as a toy.
- We live our lives distanced and distracted.

We Look for the Quick Fix

Security solutions are often implemented haphazardly. Decision makers may buy surveillance cameras or install card readers without an independent assessment or clear understanding of the real needs. That approach addresses only the symptoms, not the cause. Through advance planning and meaningful dialogue, the security professional can guide the corporate decision makers on the best long term security solution for the company.

Security professionals should take the time to ask questions and determine what the actual problem is and then create a comprehensive assets protection strategy, not a short-sighted quick fix.

INTRODUCTION TO ASSETS PROTECTION

We Both Fear and Worship Technology at the Same Time

Assets protection professionals cannot afford to be technophobes. Security systems and procedures increasingly demand an understanding of technology, and technology is becoming a major element in most business processes.

On the other hand, some people see technology as the solution to everything. Most common functions today consist of several layers of technology. If something does not work, the tendency is to add another layer of technology (Naisbitt, 2006). Careful examination of the problem might show that a solution blending technology and other solutions (training, policies, or personnel) is best.

We Blur the Distinction Between Real and Fake

The quality and quantity of electronic images (on television and in video games) tends to desensitize people to real situations. Frequently seeing people attacked or killed may make those events seem commonplace. The ramifications for security include a potential dampening of reaction by security officers and others. For example, console operators might react less quickly to events shown on their monitors because they see such things all the time in games or on television. The delay may be aggravated by information overload as security staff are expected to monitor more and more mages.

We Accept Violence as Normal

When violence is considered normal, employees may not bother to report incidents or suspicions to corporate security officials. Failure to report such matters promptly can make it more difficult to stop such situations as workplace violence, terrorism, sexual harassment, and hate crimes.

The perception of violence as normal can also affect the reaction of security officials. If they become desensitized to crime and violence, they may take incidents less seriously or react more slowly than they should.

We Love Technology as a Toy

Viewing technology as a toy can lead to a neglect of sound, risk-based assets protection strategies. For example, one company installed biometric access controls on the entrance to each of its office suites, even though there was no obvious need for high security. When asked why the equipment was installed, a manager replied, "We thought it was cool."

High technology plays an important role in assets protection, but it exacts ongoing costs, such as training and maintenance. In many situations it makes sense to step back and take a "back to basics" approach. For example, "Given a specific security challenge, imagine how you would develop a solution if you had no access to technology at all. You can then think outside the box and interject some traditional creativity into the problem-solving process" (Naisbitt, 2006).

We Live Our Lives Distanced and Distracted

Being surrounded by technology changes our relationship to other people. Assets protection professionals must never lose sight of the people factor in identifying and protecting critical assets (Naisbitt, 2006):

> Any security issue involves human psychology—and always will. The issues of safety and security are simply fundamental to every human being.
>
> When planning for security, the professionals should always consider the culture of the organization. . . . Does the corporate culture foster a sense of community? Do employees respect and care for one another? Does the nature of their work allow them to develop relationships, or do they work in a vacuum? How much human interaction is there?

In addition to the six preceding symptoms of high-tech intoxication, two other issues are worth considering:

- whether the prevalence of security technology leads employees to shirk their responsibility for protecting the organization's assets because they think technology will take care of those assets

- whether a high-tech environment depersonalizes the workplace and leads employees to feel it is acceptable to commit pilferage, industrial espionage, fraud, embezzlement, and other workplace crimes

The bottom line is that human factors must always be considered in the development of security strategies. For example, the security approach called crime prevention through environmental design (CPTED) uses psychology, architecture, and other measures to encourage desirable behavior and discourage undesirable behavior. Some critics claim that CPTED does not show a conclusive link between the design concept and a reduction in crime. However, where CPTED has been used, the recording agencies claim that there are fewer reported incidents when compared to similar structures or developments within their jurisdiction.

1.4.2 GLOBALIZATION IN BUSINESS

Globalization brings a wider range of goods, services, vendors, suppliers, capital, partners, and customers within a company's reach. It also brings threats closer and may increase vulnerabilities. Risks related to business transactions, information assets, product integrity, corporate ethics, and liability, as well as far-flung people and facilities, expand and evolve with increasing globalization. As the director of the U.S. Defense Intelligence Agency notes (Wilson, 2002):

> Values and concepts [such as] political and economic openness, democracy and individual rights, market economics, international trade, scientific rationalism, and the rule of law . . . are being carried forward on the tide of globalization—money, people, information, technology, ideas, goods and services moving around the globe at higher speeds and with fewer restrictions.
>
> Our adversaries increasingly understand this link. ... They are adept at using globalization against us—exploiting the freer flow of money, people, and technology . . . attacking the vulnerabilities presented by political and economic openness . . . and using globalization's "downsides."

Globalization makes it necessary for assets protection managers to consider a wider variety of customs, cultures, laws, business practices, economic factors, language issues, workforce characteristics, and travel requirements. A more radical vision of the impact on organizational structures is described in William Davidow and Michael Malone's *The Virtual Corporation*. They argue that the centerpiece of the new economy is a new kind of product: *the virtual product* where major business functions are outsourced with hardly any internal departmentalization. This will give the corporate security manager even more challenges in the protection of proprietary information and product security. As in all cases the dissemination of sensitive or proprietary information should be on a need to know basis. Security professionals should not erect barriers to international business but instead should help their organizations overcome those challenges and comply with the many regulations and standards that apply around the world (Heffernan, 2006).

1.4.3 STANDARDS AND REGULATION

Security standards are becoming increasingly important, and their development is the subject of much interest. The establishment of standards and guidelines has been described as the centerpiece of a comprehensive assets protection program, especially in today's global society (Dalton, 2003, p. 185). This section discusses standard-setting bodies; statutory, voluntary, and mixed standards; the use of certification and licensing as a form of standards; and the impact of regulation.

Voluntary Standards

Standards from the well-known International Organization for Standardization (ISO) are voluntary but widely adopted. Some have been integrated into various countries' regulatory frameworks. ISO standards that are relevant to assets protection involve such issues as safety and security lighting, identification cards, radio frequency identification), protection of children, and IT and information security.

In the United States, voluntary standards are also set by the National Fire Protection Association (NFPA). Many NFPA standards are incorporated into regulations, such as building codes. With the increasing integration of life safety and security systems in modern buildings, NFPA is beginning to develop security standards as well.

Several standards from Underwriters Laboratories (UL) relate to security equipment, such as locks, alarms, and access control systems. Other standards are set by trade and professional associations, such as the Illuminating Engineering Society (lighting standards and practices) and the Electronic Industries Association (electronic components and products).

In 2001 ASIS International established the Commission on Guidelines. The society has begun to issue voluntary guidelines on central security topics. The following are some of the topics addressed so far:

- general risk management
- threat advisory system response
- chief security officers
- business continuity
- private security officer selection and training
- workplace violence prevention and response
- preemployment background screening

Other guidelines are planned. ASIS guidelines are intended to increase the effectiveness and productivity of security practices and solutions and to enhance the professionalism of the industry.

Statutory or Regulatory Standards

Unlike voluntary standards, statutory or regulatory standards are binding under the law and can be enforced by formal authorities. In the United States, binding security standards are promulgated in various sources:

- Code of Federal Regulations
- National Industrial Security Program Operating Manual
- Executive Orders, Presidential Directives, and Homeland Security Policy Directives
- regulations of the Occupational Safety and Health Administration, Nuclear Regulatory Commission, Federal Energy Regulatory Commission, and Federal Trade Commission

An international source of binding standards is the International Maritime Organization.

Mixed Standards

The distinction between statutory and voluntary standards becomes blurred when voluntary standards are incorporated into laws or regulations. For example, many of the requirements in Occupational Safety and Health Administration directives are verbatim references to standards from such organizations as the NFPA.

In other situations, a standard may remain technically voluntary but practically obligatory. For example, security standards from UL or Factory Mutual may be used as criteria by insurers. In other words, they may determine the availability and cost of casualty insurance based on the use of UL-approved materials or UL-standardized practices. Contracts, too, may incorporate standards as requirements.

Figure 4 lists some of the more prominent standard-setting bodies.

FIGURE 4 *Selected Standard-Setting Bodies*	
International	
ASTM International	www.astm.org
International Electro-technical Commission	www.iec.ch
International Maritime Organization	www.imo.org
International Organization for Standardization	www.iso.org
United States	
American National Standards Institute	www.ansi.org
Department of Transportation	http://ops.dot.gov
Federal Energy Regulatory Commission	www.ferc.gov
Federal Trade Commission	www.ftc.gov
National Fire Protection Association	www.nfpa.org
National Institute for Standards and Technology	www.nist.gov
National Labor Relations Board	www.nlrb.gov
Nuclear Regulatory Commission	www.nrc.gov
Occupational Safety and Health Administration	www.osha.gov/comp-links.html
Underwriters Laboratories	www.ul.com/info/standard.htm

Professional Certifications and Licensing

Standards may also be implemented via professional certification and licensing. In the security arena, ASIS International certifications are perhaps the best-known. The Certified Protection

Professional designation, established in the 1970s, recognizes a broad skill set in security management. More recent ASIS certifications include the Physical Security Professional and Professional Certified Investigator designations.

The International Foundation for Protection Officers offers several certifications for security officers and supervisors: the Certified Protection Officer, Certified in Security Supervision and Management, and Certified Protection Officer Instructor designations.

Several IT security certifications are also available, such as the Certified Information Systems Security Professional (through the International Information Systems Security Certification Consortium) and the Certified Information Security Manager (though the Information Systems Audit and Control Association).

FIGURE 5	
Selected Security Certification Web Sites	
ASIS International	www.asisonline.org/certification/index.xml
Information Systems Audit and Control Association	www.isaca.org
International CPTED Association	www.cpted.net/certification.html
International Foundation for Protection Officers	www.ifpo.org
International Information Systems Security Certification Consortium	www.isc2.org

Specialized security certifications within particular industries are also becoming common in such sectors as health care, hospitality and lodging, and finance. Finally, certification in crime prevention is available through many state agencies and also through the International CPTED Association.

Some jurisdictions require licensing of various types of security practitioners. Most licenses require training, background screening, qualification, and registration. In the United States, licensing is generally the purview of states or localities, but national licensing is under consideration.

1.4.4 CONVERGENCE OF SECURITY SOLUTIONS

In assets protection, convergence generally means the integration of traditional and IT security functions. A broader definition might consider convergence to be the merging of disciplines, techniques, and tools from various fields for the purpose of protecting critical assets.

It is widely accepted that "companies' assets are now increasingly information-based and intangible, and even most physical assets rely heavily on information" (ASIS International, 2005). An approach using only physical or IT security measures is insufficient. Assets protection

INTRODUCTION TO ASSETS PROTECTION

managers must also employ traditional information security, personnel security, technical security, and public relations and other external communications to protect intangible assets. A true convergence approach would also employ security architecture and design, crime prevention through environmental design, investigations, policies and procedures, and awareness training.

1.4.5 HOMELAND SECURITY AND THE INTERNATIONAL SECURITY ENVIRONMENT

The terrorist attacks of September 11, 2001, made it "crystal clear that the risks and threats of global terrorism . . . were no longer vague or unlikely, but rather a genuine reality" (Sennewald, 2003, p. 19). Sennewald contends that 9/11 elevated the corporate security professional to a higher plateau of respect and recognition within the enterprise.

From an assets protection perspective, reactions to the attack have been a mixed development. On the positive side, 9/11 raised awareness of security among decision makers and increased the respect paid to the security profession. It also made resources available for security enhancements and led to increased interaction among security officials, first responders, emergency planners, and the communities they serve. On the negative side, 9/11 caused knee-jerk reactions that resulted in wasteful spending, unnecessary security measures, misdirection of needed funds, and the surfacing of dishonest or unqualified vendors.

Assets protection professionals should study those reactions and apply what they learn to comprehensive assets protection strategies. That way, they can leverage the awareness and resources available to improve their organizations' security posture.

Still, there is a danger of overemphasizing the threat of terrorism and the practice of homeland security. Assets protection professionals must address the broader security issues relevant to their particular environment.

1.5. MANAGEMENT OF ASSETS PROTECTION

In addition to technical expertise, assets protection professionals need a solid grounding in organizational management. Success in the field—which may mean saving lives and protecting valuable assets—depends on the proper balance of three managerial dimensions: technical expertise, management ability, and the ability to deal with people.

FIGURE 6
Three Managerial Dimensions

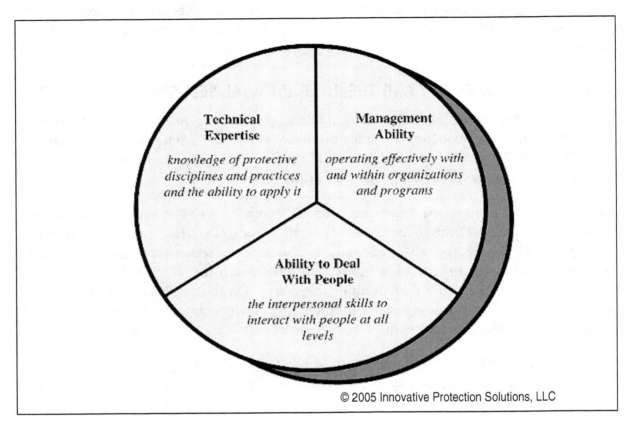

1.5.1 CONCEPTS IN ORGANIZATIONAL MANAGEMENT

The job of managing involves five basic functions:

- planning
- organizing
- directing
- coordinating
- controlling

In addition, management should be guided by two principles, called "who is the customer?" and "quality." These principles should become part of the organization's culture.

Who Is the Customer?

Peter Drucker, an authority on management, suggests that "who is the customer?" is the first and most crucial question in defining business purpose and mission (1974). The assets

INTRODUCTION TO ASSETS PROTECTION

protection manager must understand the purpose and mission of assets protection at the enterprise before adopting an organizational structure.

Most organizations actually serve multiple customers. It is important to identify all of them and to understand their interrelationships. Then the assets protection manager can sell the program not just to executives but to all the customers of assets protection services. Figure 7 lists some of those customers.

FIGURE 7		
Assets Protection Customers		
For a chief security officer or security director, customers might include:	**For a security product or service provider, customers might include:**	**For an independent consultant, customers might include:**
Corporate executives	Clients	Clients
Corporate staff/managers	Clients' clients	Clients' clients
Corporate employees	Potential clients	Potential clients
Company clients	Parent company or headquarters	Partners and associates
Partners and affiliates	Vendors and suppliers	Vendors and suppliers
Contractors	Partners and consultants	Own employees
Security team members	Original equipment manufacturers	Investors
Vendors and suppliers	Own employees	Self
Other divisions of company	Other divisions of company	
Other facility users	Executive management	
Stockholders	Stockholders	

Taking a more comprehensive view of who the customers are and how best to meet their needs can result in greater security team effectiveness. The large view also demonstrates the assets protection manager's commitment to the business mission as a whole, not just to the security mission. That commitment often leads to greater respect for the assets protection function and ultimately greater influence throughout the enterprise.

Quality

Some managers may think that quality is something in a plan on the shelf, something that is done once, or something that belongs to the quality assurance experts. That view is wrong. Quality "belongs to everyone, all the time" (Dalton, 2003, p. 240).

As one quality consultant notes (Duffy, 2006):

One of the major definitions of quality is "conformance to customer requirements." Providing effective professional services or implementing a meaningful assets protection program for the customer within appropriate resource constraints means delivering the required level of quality. The security industry is one that must support multiple customers with a wide variety of requirements.

Although a quality program may begin with tools, measures (metrics) and special processes, the culture of quality should ideally become a part of the organization and be integrated into all business practices.

A culture of quality can be developed in any type of security organization. For example, security service providers are increasingly formalizing and standardizing their quality programs.

1.5.2 MANAGEMENT APPLICATIONS IN ASSETS PROTECTION

Planning, management, and evaluation are important tools in crime prevention programs (Fennelly, 2004, p. 418). A strategic approach to managing assets protection programs likewise involves all three tools. They apply as follows:

- **Planning** includes developing strategic goals and objectives, aligning assets protection objectives with the organizational vision, organizing the assets protection function in the way that best meets objectives, and determining how the mission will be accomplished.

- **Management** involves conducting the day-to-day operations of the department, communicating with others, and controlling specific tasks as well as the overall functioning of the office.

- **Evaluation** involves stepping back from day-to-day activities to objectively assess how well objectives are being met and what factors are contributing to the success or lack thereof. Reporting, documenting, and using information to make adjustments and improvements are all important parts of evaluation.

These tools as applicable in the security services or products arena as they are in the corporate or organizational setting. In a quality assurance/quality control (QA/QC) program in a firm that provides security officers, the tools could work as follows:

- **Planning** may entail developing the company's QA/QC program, obtaining executive buy-in, preparing documentation, training supervisors, and establishing procedures.

- **Management** might involve implementing the program, conducting inspections, reviewing audit reports, handling complaints and compliments, disciplining and rewarding officers and supervisors, briefing upper management, and interacting with the client on matters pertaining to QA/QC.

INTRODUCTION TO ASSETS PROTECTION

- **Evaluation** could consist of periodically determining whether the QA/QC program is serving company objectives and meeting client expectations, identifying systemic problems, and recommending process improvements.

In a corporate setting, a security department could use the tools as follows:

- **Planning** may entail setting strategic objectives consistent with the enterprise's mission and vision statements, organizing the security function within the enterprise, determining resource requirements, establishing liaison relationships, developing policies and procedures, and identifying staffing needs.

- **Management** would involve day-to-day operation of the department, personnel management, logistics, vendor management, security systems operations, coordinating with others internally and externally, and briefing senior executives.

- **Evaluation** would consist of periodically comparing performance metrics to the department's goals and objectives, identifying shortfalls, assessing any changes in the assets protection environment, and recommending process improvements.

None of these functions should be neglected at the expense of the others. They should be repeated in an ongoing cycle that results in up-to-date and appropriate assets protection protocols, procedures, and practices.

1.5.3 SECURITY ORGANIZATION WITHIN THE ENTERPRISE

Although each organization is unique, some basic principles apply widely to organizational structure and management. This discussion of the security organization within an enterprise is influenced by well-respected, much recommended security textbooks by Sennewald (2003), Dalton (2003), McCrie (2001), and Fischer & Green (2004).

The "span of control" principle suggests that a single person can supervise only a limited number of staff members effectively. The specific number depends on such factors as the nature of the work and type of organization, but as a general rule one manager can effectively supervise up to 10 people. This principle may be in jeopardy. Some observers believe that the introduction of IT infrastructures, use of current telecommunications technology, and flattening of organizational structures may enable a person to supervise as many as 100 people. In settings that emphasize self-directed, cross-functional teams and very flat structures, span of control is less relevant. However, traditional, hierarchical organizational structures, where span of control is important, are still common.

Unity of command dictates that an individual report to only one supervisor. It is based on the concept that a person cannot effectively serve the interests two or more masters (that is, managers). It is the supervisor's responsibility to ensure the best performance from the unit they manage. Some company structures make unity of command less important, but in most settings employees still need a clear understanding of which policies they need to adhere to (primarily) and who will provide day-to-day direction, quality control, and conflict resolution.

Placement of the security department within an organizational structure can greatly affect the assets protection manager's ability to exert influence, remain informed, and garner resources to support his or her programs and strategies. Asset protection managers, by the nature of their expertise, must have functional authority within the organization and be identified as part of the corporate management team. The rule of thumb is that the senior security or assets protection professional should be placed as high as possible in the structure of an enterprise, and report directly to senior or executive management. A common discussion today is whether security should be placed under the chief information officer), IT security should be placed under a chief security officer, or some other arrangement should be made. If the enterprise includes a chief risk officer, assets protection may be placed in his or her division.

The following are some other important themes in organizational management:

- Lines of authority, responsibility, and communications should be as clear and direct as possible.

- Individual and organizational responsibility should come with an appropriate level of authority.

- Organizational alignments and structures should consider the inter-relationships among functions, roles, and responsibilities (with an eye on the overall mission).

- Communications channels should be structured to allow effective mission accomplishment and interaction.

More information on the chief security officer's role in organizational management can be found in the *Chief Security Officer Guideline,* published by ASIS International (2004). It discusses roles and responsibilities, success factors, key competencies, organizational issues and strategy development.

1.6. BEHAVIORAL ISSUES IN ASSETS PROTECTION

Behavioral science, the study of people and their relationships to each other, is important in assets protection for two key reasons:

- Many security risks are the result of human threats, and behavioral science can yield insights into human threat sources.

INTRODUCTION TO ASSETS PROTECTION

- Security management requires effective interaction with other people, including collaboration, education, influence, supervision, and the most important, excellent communication skills.

- An effective security manager must also have trust in his staff and have the ability to delegate not only the responsibility, but also the authority to act within their functional area.

1.6.1 BEHAVIORAL SCIENCE THEORIES IN MANAGEMENT

The following theories in behavioral science are widely accepted as relevant and useful in many management applications.

Maslow's Hierarchy of Needs

Abraham Maslow's theory, commonly known as the hierarchy of needs, asserts that people's behavior is driven by basic needs at different levels. It is often depicted as a pyramid, as Figure 8 shows.

FIGURE 8
Maslow's Hierarchy of Needs

The levels of the hierarchy are:

- self-actualization need: self-fulfillment, realizing one's full potential
- esteem or recognition needs: respect from others and self
- affiliation or love needs: affectionate social and family relationships
- security or safety needs: protection from perceived harm
- physiological or survival needs: food, drink, shelter

Basic or lower-level needs must be met before a person is motivated by the next higher level of needs.

Maslow's theory is still widely recommended to analyze individual employee motivation strategies and establish tailored rewards, such as pay, recognition, advancement, and time off (Buhler, 2003).

McGregor's Theory X and Theory Y

Douglas McGregor holds that two worker models can be contrasted. Theory X contends that workers are inherently lazy and tend to avoid work. They lack creative ambition, must be goaded, require constant supervision, and are motivated by fear. Theory Y states that workers are naturally motivated and want to work hard and do a good job. It assumes that workers are thoughtful, eager to perform well, and willing to be guided and taught. McGregor stresses that programs based on Theory Y are more successful than those based on Theory X.

Herzberg's Motivation-Hygiene Theory

Frederick Herzberg's motivation-hygiene theory is based on the premise that the opposite of satisfaction is not dissatisfaction but simply no satisfaction. The theory maintains that two sets of factors determine a workers' motivation, attitude, and success (Buhler, 2003).

The first set is job content (motivators), such as achievement, recognition, responsibility, and satisfaction derived from the work itself.

The second set is job context (hygienes), such as the surroundings, physical work conditions, salary, coworkers, and other factors that are external to the work itself.

Hygiene factors (such as a fresh coat of paint on the wall) will be able to move an individual from a state of dissatisfaction to no satisfaction, but only motivation factors can move that person from no satisfaction to satisfaction.

The lesson is that managers should avoid quick fixes. Manipulating hygiene factors may alleviate dissatisfaction but will not result in a state of satisfaction. Allowing an individual to reach a state of satisfaction requires changes in the work content itself, such as increased autonomy or responsibility (Buhler, 2003).

1.6.2 APPLICATIONS OF BEHAVIORAL STUDIES IN ASSETS PROTECTION

An assets protection program will not succeed unless it cultivates the willing cooperation of those affected by it and meshes its goals with the personal goals of the workforce. Following are some examples of how lessons from behavioral science might be employed in assets protection.

Crime Prevention and Reaction

Behavioral science has long been involved in criminology with the goal of developing better crime prevention strategies. Through mutual cooperation, private security can play a major role in the prevention of crime while law enforcement focus on crime control. Continuing study is needed, as is better communication between behavioral scientists, criminologists, and security and law enforcement practitioners. Many questions in criminology remain unanswered in this area, but we are seeing a major move by law enforcement to have private security more involved in crime prevention.

Incident Management

Motivation theories may be useful in developing emergency plans, business continuity plans, and incident response plans. A major factor in any incident is how people will react—those directly involved in the incident, bystanders, indirectly affected persons, security forces, and first responders.

Some data can be gathered from exercises and drills through documentation and after-action reports. Interpreted through human motivation theories, that information may aid in the development of plans and procedures that will help ensure a smooth response to a real incident.

Motivation theories should also be considered when developing larger-scale incident management plans. Such theories may help in predicting how people will react when they are ordered to shelter in place at the workplace or school—for example, whether they will accept their separation from their family or instead evacuate immediately, regardless of the directions given.

Security Personnel Management

In supervising security officers, heading an executive protection team, staffing a security operations center, serving as a facility security officer, performing architecture and design functions, or administering a global assets protection program, one needs to understand what motivates people and what demotivates them.

Motivation theory can contribute to the planning and development of a QA/QC program, a department organizational structure, an advancement plan, assessment or evaluation criteria, awards programs, discipline procedures, communications venues and even dress codes. Behavioral science plays a role in almost every aspect of personnel management.

Employee Training and Awareness

Early security training and awareness programs were based on top-down management directives, passive compliance, and an attitude of "we do it this way because the book says we do it this way." The modern workforce is more sophisticated, highly educated, and independent, and security training and awareness strategies must be designed accordingly.

Behavioral theories can guide both content and delivery methods for security training and awareness, which has been recognized as one of the most cost-effective assets protection tools (Webster University, 2006). In addition, security training and awareness efforts should take account of adult learning styles and current instructional design methods. When employees can relate to the information presented and the way it is presented, the training is more effective. Managers need to set direction and establish a professional setting, but through training they need to avoid making operating decisions that should be made by their supervisors and officers. As an example, when a subordinate requests advice about a routine operational problem, the supervisor should avoid giving a specific solution, opting instead to guide the subordinate, through an open exchange of information, to identifying the solution him or herself.

Corporate Ethics

One of the first questions that comes to mind after a large-scale corporate scandal is "What could have possibly motivated those people to do that?" Behavioral science theories may help answer that question. They can be applied to help prevent, respond to, and recover from major white-collar crime incidents and can also contribute to programs that address smaller-scale, everyday ethical lapses.

Liaison and Leveraging Other Organizations

Because assets protection is a multidisciplinary venture, liaison and collaboration with a wide variety of people, organizations, agencies, specialties, and professions is essential. Behavioral theory can help in establishing and maintaining relationships with a network of professional contacts, both inside and outside the assets protection manager's organization.

Collaboration is especially valuable and challenging in a global environment that includes a wide range of cultures, customs, and perspectives (Buhler, 2003):

> The diversity of today's workforce has further complicated an already complex phenomenon. The differences among workers are greater than ever before. To be more successful in motivating a diverse workforce requires, then, an understanding of the differences among people and what makes them tick. ...

> To become a more effective motivator, then, managers must understand as much as possible about [motivation theory] and then pick and choose what best fits with which individuals. The bigger the bag of motivational tools, the more likely the manager will be able to understand employees' needs and tailor rewards to better meet them. [This] enables managers to get more done through others.

APPENDIX A

INSURANCE AS A RISK MANAGEMENT TOOL

In many organizations, a current trend is the integration of insurance management into a broader assets protection program. Therefore, this appendix describes the types and uses of insurance, primarily in the corporate setting. Further information is available through resources listed at the end.

Most risk management tools are either proactive or reactive, but insurance is a combination of the two. From a proactive stance, it is the best-known form of risk transfer and is actually considered an asset of the organization. It is also reactive in that the insurance benefits are not used until after a loss occurs.

Insurance is a formal undertaking between two parties—the insurer and the insured—under which the insurer agrees to indemnify or compensate the insured for specified losses from specified perils. Insurance is "a formal social device for reducing risk by transferring the risks of several individual entities to an insurer. The insurer agrees, for a consideration, to assume, to a specified extent, the losses suffered by the insured."[1]

Insurance is no replacement for security, of course. Compared to insurance, protection techniques like risk reduction and risk spreading are preferable for several reasons:

- Loss control is a more satisfactory approach than after-the-fact indemnity.
- Loss prevention has become highly effective.
- Commercial insurers decline to cover some kinds of risks.
- The balanced scheme of protection is more cost-effective.

In most cases, it is impossible to be fully compensated for a loss, regardless of how much insurance coverage an enterprise has. Modern management is now more interested in preventing losses than in trying to buy insurance to cover every possible risk.

In the insurance world, the portfolio theory involves a comprehensive analysis of business risks and pure risks. A risk model might analyze movements in exchange rates, changes in raw material prices, and downtime caused by a catastrophic event. This model would produce an aggregate loss distribution to estimate the likelihood and effect of several events occurring simultaneously. By treating the risks as parts of a single portfolio, separate insurance policies

---

[1] *Glossary of Insurance Terms,* University of Calgary, Canada, 1998, http://www.ucalgary.ca/MG/inrm/glossary/index.

for each risk can be eliminated. The theory is that by managing risks, little or no outside insurance is required.

INSURANCE OVERVIEW

Insurance is often divided into two general categories: property and liability. Property coverage includes building and equipment damage or loss, as well as items like cash and negotiable instruments of all kinds. Liability coverage encompasses all employee risks and includes workers' compensation and non-occupational coverage, as well as coverage for losses affecting the general public, such as automobile liability, product liability, landlord liability, contractor liability, and environmental liability.

The basis for coverage is the insurance policy, the written contract between the insurer and the insured. Many insurance contracts or policies have been standardized; however, they are not all alike in coverage. For that reason, each policy must be carefully examined to determine the coverage offered. *Contracts of insurance are seldom read in detail by the owners until a loss occurs.* To determine the protection offered by a policy, the following questions must be asked:

- What perils are covered?
- What property is covered?
- What losses are covered?
- What people are covered?
- What locations are covered?
- What time period is covered?
- What hazards are excluded or what conditions suspend coverage?

Defining the Peril

Peril has been defined as "the cause of a possible loss."[2] Typical insurable perils include fire, windstorm, explosion, burglary, negligence, collision, disability, and death. An insurance contract may cover one or more perils. Some policies, called "named perils contracts," specify the perils that are covered in the contract. Other contracts, called "all risk contracts," cover all perils except those that are specifically excluded. Perils may also be covered only in part—for example, not all unfriendly fires under a fire policy or not all negligence under a liability policy.

---

[2] *Glossary of Insurance Terms*

A policy may limit coverage by defining which part of the peril is covered or which part is not covered. For example, a fire policy states the hazards not covered. The standard policy form excludes fire losses resulting from action taken by military, naval, or air forces in an actual or immediately impending enemy attack, invasion, insurrection, rebellion, revolution, civil war, or usurped power. It also excludes fire losses resulting from neglect of the insured to use reasonable means to protect his or her property, along with losses caused by order of civil authority (except destruction of property to prevent the spread of a fire that did not originate from an excluded peril).

It is important to understand the terms burglary and robbery as they are used in insurance policies. Burglary is generally defined as felonious abstraction of insured property by any individual or individuals gaining entry to the premises by force.[3] There must be visible marks on the exterior of the premises at the place of entry, such as evidence of the use of tools, explosives, electricity, or chemicals.

Robbery is usually defined as the felonious and forcible taking of property by violence inflicted upon a custodian or messenger, either by putting the person in fear of violence or by an overt act committed against the custodian or messenger who was cognizant of the act. Sneak thievery, pickpocketing, confidence games, and other forms of swindling are not included in robbery coverage.

A burglary contract does not cover robbery. Similarly, a robbery policy does not cover burglary. Neither policy covers losses resulting from the felonious taking of property where there are no visible marks of entry and where there has been no violence or threat of violence. A theft or larceny policy is required to obtain coverage for such losses.

Defining the Property Covered

A standard insurance policy does not cover every piece of property owned by the insured, but it usually describes the type of property covered. Also, a contract may specify certain property that is excluded.

Some reasons for property exclusions in a policy are as follows:

- The specific property excluded may be more easily covered under other forms of insurance.

- The moral hazard—a condition of the insured's personal habits that increases the probability of loss—may be prohibitive.

---

[3] In law, burglary is forced entry or exit with intention to commit a crime. The abstraction of property is actually a larceny. But insurance policies combine the forceful entry and the taking or abstraction under the single term burglary.

- The property may be subjected to hazards that should be specially rated.
- The property might be so uncommon to the average insured that the rate for the standard policy should not include it.

Defining the Losses Covered

The next step in analyzing coverage is to find out what losses are covered. Generally, losses may be classified as:

- **direct loss,** such as the physical loss of or damage to the object concerned
- **loss of use,** such as the reduction of net income due to loss of use of the damaged or destroyed object
- **extra-expense losses,** such as the costs of defending a liability suit and paying judgment or hospital and medical expenses following a personal accident

Most policies cover direct losses only. Some may, in addition, cover a few forms of indirect losses. For example, a standard fire insurance policy usually covers only the actual cash value of the property at the time of the loss. Actual cash value is the cost to replace or restore the property at prices prevailing at the time and place of the loss, less depreciation.[4] It will not offer compensation for additional expenses of rebuilding required by ordinances regulating construction or repair, and it will not cover the loss of use while the property is being replaced. In addition, it will not pay for the loss of income, such as loss of rent, while a building is being rebuilt.

Defining the Period of Coverage

Formerly, a loss that occurred during the period the policy was in force would be covered no matter when the occurrence was discovered, even after the policy expired. The term for this is an occurrence loss.

Insurance carriers encountered difficulties matching premiums with losses that could still be covered years after occurrence. As a result, a new form of contract was developed. This form, known as the claims-made type, provides coverage only for losses that are reported during the period the policy is in force.

If an insured with a claims-made policy leaves one carrier in favor of another, the new carrier will probably not cover losses occurring before its own first contract date, even if the claim is made during the contract period. This tends to lock insureds in with a single carrier. It also raises issues of later endorsements to reduce coverage, the need for an insured to solicit

---

[4] *Glossary of Insurance Terms.*

claims against itself in order to pass them to the carrier in a timely way, and the uncertainty of coverage or its cost when seeking to terminate the contract. The solution to this problem is usually called "tail cover"—retrospective coverage for events that occurred during a prior policy period but are raised during the tail period. To change carriers, it is normally necessary to purchase tail cover from the prior carrier.

Defining the People Covered

Some policies cover only the named insured and representatives while others cover additional individuals. The first page of a standard fire policy states clearly that the contract insures only the named insured or insureds and legal representatives. The insured's executors or heirs under a will and receivers in bankruptcy would also be covered. Many property policies allow a space for indicating the name of the lender who holds a financial interest in the property, and such lenders are considered additional insureds. An endorsement must be added to afford protection to any others. A frequent technique to extend one party's coverage to protect another is to have the other individual designated as a named insured in the policy. Named insureds, however, are subject to the same policy conditions as the original insured. In some cases, this may not achieve the security objective of the additional named insured.

Defining the Locations Covered

Some policies cover one location, while others include several locations. The standard fire insurance contract covers property only while it is located as described in the policy, with one exception—the contract covers property pro rata for five days at each proper place to which any of the property is necessarily removed for protection against the perils insured against in the policy.

Defining the Time of Coverage

Policies vary as to the exact time of day they go into effect. Fire insurance policy coverage usually starts at noon, standard time, on the day the policy is dated and at the place the risk is located. The coverage will ordinarily continue in force until noon, standard time, on the day of expiration. Other policies go into effect and expire at 12:01 a.m., standard time.

Conditions that Suspend Coverage (Exclusions)

Insurance policies commonly contain provisions that suspend coverage when a risk increases to such a degree that the insurance company is no longer willing to offer protection. It is possible to eliminate the conditions by adding endorsements, which may result in increased premiums.

The limiting provisions may be either "while" clauses or "if" clauses. That is, coverage is suspended *while* certain conditions exist or *if* defined situations exist. The fraud and concealment clause found in many contracts is a typical "if" clause. It states that coverage is void if, either before or after a loss, any material fact or circumstance concerning the insurance has been willfully concealed or misrepresented. An example of a "while" clause would be a statement that the insurance company will not be liable for loss while the hazard is increased by any method within the control or knowledge of the insured. Another common example would be the vacancy clause, which suspends coverage while a property stands vacant beyond a specified period.

In fidelity coverage, it is customary to exclude from coverage any person the insured knows to have committed any fraudulent or dishonest act, in the insured's service or otherwise. The exclusion usually dates from the time the insured became aware of the fraudulent or dishonest act. The insurance carrier may grant case-by-case exemptions to the exclusion. For example, should a person be hired despite a minor dishonest act revealed in a preemployment investigation, an exemption to the exclusion should be requested.

Endorsements

Insurance policies have been standardized by custom, law, or inter-company agreements. Standard policies may be modified by endorsements—sometimes called riders—to increase or decrease the coverage of the standard policy. Standard endorsements are available, but if they are not adequate for the coverage desired, special endorsements may be written and added to the standard policy. When in conflict with the standard policy, the endorsement governs unless it is illegal.

Endorsements are added to:

- add perils
- add property
- include more covered individuals
- adjust rates
- add, increase, reduce, or delete deductibles
- add or eliminate exclusions
- increase or decrease amounts of coverage
- record address changes
- correct errors

Crime Coverage

Crime insurance is written to protect the insured against loss by burglary, robbery, theft, forgery, embezzlement, and other dishonest acts. Two types of bonds may be used for protection: fidelity and surety. Fidelity coverage is written to protect the employer from the dishonesty of employees. Surety coverage is intended to guarantee the credit or performance of some obligation by an individual.

Insurance coverage against crime may be obtained by purchasing a standard crime policy, then adding the necessary endorsements. It is essential to understand the meaning of each criminal term used by the insurance company in order to ensure that adequate protection is obtained. Policies may exclude certain items or may not include certain crimes.

The comprehensive 3D policy is a combination fidelity crime insurance policy designed to offer the widest possible protection. The standard form contains five insuring agreements. The insured may select as many as needed and specify the amount of coverage on each. The following are the basic coverages offered:

- Coverage I—an employee dishonesty bond
- Coverage II—money and securities coverage inside the premises
- Coverage III—money and securities coverage outside the premises
- Coverage IV—money order and counterfeit paper currency coverage
- Coverage V—depositors' forgery coverage

Twelve additional endorsements are available:

- incoming check forgery
- burglary coverage on merchandise
- paymaster robbery coverage inside and outside premises
- broad-form payroll inside and outside premises
- broad-form inside premises only
- burglary and theft coverage on merchandise
- forgery of warehouse receipts
- securities of lessees of safe-deposit box coverage
- burglary coverage on office equipment
- theft coverage on office equipment

- paymaster robbery coverage inside premises
- credit card forgery

Assets protection managers should consider an endorsement for IT equipment and data if they are not adequately covered in the policy. In determining whether coverage is adequate, the following questions should be asked:

- Is all equipment completely covered for any loss?
- Does the coverage include the loss of recorded data as well as the cost of new hardware?
- Does the coverage include reconstruction of data?
- Will the coverage pay for temporary operation at an alternate location?
- Does business interruption coverage protect against forced shutdown of equipment?

Business Interruption

Business interruption insurance offers a number of coverage choices. For example, coverage can be written on a named peril or all risk basis. If a building or machine sustains physical damage, there will usually be at least an interruption of production or sales, resulting in financial loss. Other incidents may not damage the physical facilities but may nevertheless cause a shutdown. For example, a subcontractor might be required to shut down if the plant of the prime contractor is destroyed, or a factory across from a chemical plant might be forced to lose a day's production because of noxious fumes from the chemical plant. These types of risks can be covered with endorsements known as contingent business interruption loss forms.

A business that might not return to normal for some time after reopening following a shutdown could consider another type of coverage: the endorsement extending the period of indemnity. An example of a business requiring such coverage would be a bowling alley. A fire just prior to the opening of a bowling season might cause league business to go elsewhere for the full season. Even if the establishment is able to reopen in two months, it might not recover its normal business until the following year. With standard business interruption insurance, the coverage would stop once the facility was restored to operating condition. With the endorsement extending the period of indemnity, the coverage would be extended for the amount of additional time purchased.

Valuation is a factor to consider in planning for business interruption. An actual-loss-sustained method or a valued-loss method may be selected. With actual-loss coverage, the insured must prove the claim according to policy provisions. On the other hand, the valued endorsement usually stipulates the amount payable per day of shutdown and specifies the number of days for which coverage is provided. The amount selected for the daily indemnity must be certified by an accountant as being the approximate amount that will actually be lost. This certification is done before the loss occurs.

Another type of business interruption insurance is the business interruption and extra expense endorsement. While the basic business interruption forms include coverage for normal extra expenses, other expenses may be incurred. Such expenses may be incurred to keep a product on the market regardless of cost or, for a bank, to function regardless of expense. When the situation is not a clear-cut case of either loss of earnings or incurring extra expense, a combined endorsement may offer good protection.

Liability Endorsements

Liability coverage in recent years has become increasingly important because of cases in which organizations have been held liable for property damage and for injury to victims. Under tort law, injury victims are entitled to collect for losses and mental anguish from anyone they can prove responsible for intentionally or negligently injuring them or damaging their property.

Liability litigation is widespread, and the number of liability cases continues to rise. Products are challenged as unsafe or badly designed, and such actions frequently result in large damage awards. Professional liability suits against engineers, architects, physicians, and lawyers have multiplied, and the cost of liability insurance for some professionals is enough to cause them to abandon their practice.

In the security field, too, liability litigation has exploded, resulting in many large damage awards against security personnel, contract security agencies, and employers or client companies.

A commercial general liability policy—the standard policy offering liability coverage—is less comprehensive than generally assumed. As a result, to ensure the necessary coverage, several endorsements should be added, such as those below.

Liability of Officers and Directors

A liability endorsement to protect officers and directors against legal actions brought by stockholders and others has become increasingly popular because of the publicity given to such suits. Coverage should be carefully examined to ensure that it is adequate. For example, a policy may specify that protection is offered for individuals "while acting within the scope of their duties." This provision could lead to questions as to duties of individuals and whether they were acting within the scope of those duties. An endorsement providing for coverage while "acting in behalf" of the enterprise would eliminate such a dispute. Such a change can usually be made without any increase in premium.

Employee Practices Liability Insurance (EPLI)

This relatively new type of insurance is a specialized coverage for employers who become the targets of work-related lawsuits. EPLI covers a business for employee-related actions, such as the following:

- discrimination
- sexual harassment
- wrongful termination
- breach of employment contract
- negligent evaluation
- failure to employ or promote
- wrongful discipline
- deprivation of career opportunity
- wrongful infliction of emotional distress
- mismanagement of employee benefit plans

EPLI covers defense costs, judgments, and settlements but may not cover punitive damages, fines, or penalties. Workers' compensation, bodily injury, and property damage, and any liability covered specifically in another policy are generally not covered. EPLI usually covers the corporate entity, employees, former employees, directors, and officers. Some policies also cover volunteers.

Product Liability

Product liability insurance is sold to manufacturers and dealers of goods. Protection is offered for damage claims arising from the consumption or use of articles manufactured, sold, handled, or distributed by the insured, if the damage occurs after possession of the goods or products has been relinquished to others and if the damage occurs away from the insured's premises. An exception exists for organizations that serve food on the premises, for which special coverage is necessary.

Product liability suits may be based on either the tort theory of negligence or the contract theory of breach of warranty. Since it is easier to prove breach of warranty than negligence, most claims involving products are based on a breach of an express warranty or an implied warranty that the product sold is reasonably fit for the particular purpose for which it was bought. Liability coverage must be examined carefully to ensure that breach of warranty is included. If not, an endorsement should be added for this protection.

The recall of products, which is excluded in standard liability coverage, can create an expensive problem. Frequently, manufacturers are required to recall automobiles, television sets, food products, or pharmaceuticals. The manufacturer is normally required to assume responsibility for removing the defective item from the possession of all wholesalers and retailers.

Product recall coverage can be obtained by adding an endorsement to the comprehensive liability policy. This coverage is known as product recall or product withdrawal expense. The coverage may be written to cover recall of products only if bodily harm is threatened, or it may cover products that threaten only property damage. The loss of the product itself is not covered.

Insurance Providers

Regardless of the type of insurance provider, customers should be able to expect rapid compensation for losses incurred. As in any other business relationship, due diligence must be exercised when selecting an insurance provider. The financial stability and claims settlement record of the provider is critical to timely reimbursement of a loss. Most organizations select an insurance provider and settle into a long-term business relationship without subsequent review of the financial condition of the provider, but ongoing due diligence is necessary.

Insurance can be obtained through these means:

- dealing directly with an insurance company
- dealing with an insurance broker that may represent several companies
- buying an insurance company, known as a captive carrier
- buying an interest in a mutual insurance organization called a risk retention group

The size of the enterprise and its insurance needs typically suggest the type of provider that will be most cost-effective. Small organizations tend to deal directly with the insurance company or use a broker. Mid-size organizations have the same options but may also join a risk retention group. Large organizations have all four of the options listed above. The four different sources of insurance are discussed below.

Insurance Companies

The large number of insurance companies and the wide variety of policies they offer ensures that coverage can be found for virtually any risk. In essence, uninsurable risk is only heretofore uninsured risk. Many organizations merely select an insurance carrier with a good name, accept the coverage that the representative suggests, and pay the policy premiums. Sound management principles demand more.

A financially weak carrier tends not to pay claims in a timely manner. If the carrier becomes insolvent, claimants can turn to the state guarantee trust fund for partial recovery. This is a lengthy process, and claimants are limited to a certain dollar amount. In essence, choosing the wrong insurance company can, in itself, be a high risk. The financial stability of the insurance carrier should be reviewed before entering a contractual relationship, and subsequent reviews should be conducted at least annually.

The financial stability of insurance carriers is rated by a number of rating services. Each service uses a different formula, and the rating of a specific insurance company may vary among the rating services. Prudent managers consult more than one rating service. A significant difference in the ratings of a company should be a red flag denoting the need for further investigation. Rating services measure the financial condition of the insurance carrier but do not measure the speed of claims payments.

Government insurance departments are also valuable sources of information. In the United States, in each state insurance companies are authorized to do business in, they must file annual financial statements with the state insurance department. Other pertinent information includes the number of complaints filed against the company and any disciplinary action taken against the company.

Insurance Brokers

Insurance brokers are marketing specialists who represent buyers of property and liability insurance and who deal with either agents or companies in arranging for the coverage required by the customer.[5] Insurance brokers deal with more than one insurance company and can suggest the company best suited to provide a specific type of policy. The expertise and responsiveness of a broker should be verified by contacting other clients. A good broker keeps abreast of the financial stability of the insurance companies with which insurance is placed. The broker who arranges insurance coverage with an insurance company that becomes insolvent may become a defendant in a civil action.

Risk Retention Groups

Smaller firms and organizations may form risk retention groups (RRGs), which are corporate bodies authorized under the laws of some states as liability insurance companies. Such groups

---

[5] *Glossary of Insurance Terms.*

must be owned by entities within the membership of the group that obtain liability insurance from the group. RRGs are generally exempt from the laws of other states.

RRGs typically market their liability policies to purchasing groups (PGs), which consist of organizations that have similar liability insurance needs because of the nature of their business. In the security field, PGs have consisted of guard and investigations concerns. The PG can acquire liability insurance for its members from the RRG. Typically, the attraction of such an approach has been the availability of liability coverage and lower premiums. Some RRGs have experienced funding or other difficulties and have either abandoned the field or otherwise caused problems for the PG insureds. Overall, the RRG is a viable alternative to high premiums and the difficulty of obtaining special coverage; however, the particular group and its track record should be studied carefully.

Captive Carriers

One of the problems of liability insurance has been the high premium cost when using carriers conventionally licensed within each state where they offer the coverage. One solution is the captive insurer—a separate, wholly or principally owned firm, usually organized offshore, used to write the insurance for the owning company. Sometimes a captive insurer is owned by an association of two or more firms with common insuring interests. When appropriate, a captive insurance carrier can make it easier to insure risks not acceptable to conventional carriers, can help make a more favorable expense ratio, and can open reinsurance resources not otherwise available. However, the captive carrier is generally a technique of larger firms.

INSURANCE RESOURCES

Business Insurance magazine and online resources
www.businessinsurance.com

Insurance Information Institute
www.iii.org

Risk Insurance and Management Society
www.rims.org

"The Smart Approach to Protecting Your Business: Managing Your Risk"
The Hartford in association with the U.S. Small Business Administration
www.sba.gov/library/pubs/mp-28.pdf

REFERENCES

ASIS International. (2004). *Chief Security Officer Guideline.* Alexandria, VA: ASIS International.

ASIS International. (2005). *Scope and emerging trends: Executive summary.* Alexandria, VA: ASIS International.

Buhler, P. M. (2003, December). Managing in the new millennium: Understanding the manager's motivational tool bag. *Supervision.*

Dalton, D. R. (2003). *Rethinking corporate security in the post 9/11 era.* Burlington, MA: Butterworth-Heinemann.

Drucker, P. F. (1974). *Management tasks, responsibilities, practices.* New York, NY: Harper and Row.

Duffy, G. (2006, September 23). Vice President, American Society for Quality, www.asq.org. Unpublished Document.

Fennelly, L. J. (2004). *Handbook of loss prevention and crime prevention* (4th ed.). Burlington, MA: Elsevier Butterworth-Heinemann.

Fischer, R. J., & Green, G. (2004). *Introduction to security* (7th ed.). Burlington, MA: Butterworth-Heinemann.

Glassman, C. A.. (2006, June 8). Complexity in financial reporting and disclosure regulation. Presentation at the Security and Exchange Commission and Financial Reporting Institute Conference, Pasadena, CA.

Heffernan, R. J., CPP. (2006, September 25). 2006 trends in proprietary information loss survey results: An overview. Presentation at the ASIS International Seminar & Exhibits, San Diego, CA.

McCrie, R. D. (2001). *Security operations management.* Burlington, MA: Butterworth-Heinemann.

Naisbitt, J., Naisbit, N., & Phillips, D. (1999). *High tech/high touch.* New York, NY: Broadway Books.

Naisbitt, N. (2006, June 22). Founder and executive director, The Pinhead Institute, Telluride, CO. Personal interview.

National Institute for Standards and Technology, Computer Security Resource Center. (2006). History of computer security. Available: http://csrc.nist.gov/publications/history [2006, July 28].

Securitas. (2006). History. Available: http://www.pinkertons.com [2006, July 28].

Sennewald, C. A., CPP. (2003). *Effective security management* (4th ed.). Burlington, MA: Butterworth-Heinemann.

Webster University. (2006). *Business assets protection.* Course materials for Business and Organizational Security Management Program. Washington, DC: Webster University.

Wilson, T. R. (2002). Global threats and challenges. Statement to the U.S. Senate Armed Services Committee by the Director of the Defense Intelligence Agency, March 19, 2002.

PART 2
Closed-Circuit Television Systems

This text is excerpted from the March 2007 revision, published as "Part III: Closed-Circuit Television Systems" in the "Other Hardware" chapter of the ***Protection of Assets (POA) Manual***.

2.1. INTRODUCTION

In recent years, closed-circuit television (CCTV) technology has improved greatly. The transition from analog to digital technology has changed the foundations of system design. In fact, the fast pace of development may lead a security manager to wonder how much new information he or she must master to stay current with CCTV trends. As the technology becomes better, smaller, and more reliable, the number of applications increases, but fortunately the theory of CCTV application remains the same.[1] This part of the *Protection of Assets Manual* discusses the theory of designing CCTV security systems, the changing technology, and guidelines for choosing equipment.

In designing a CCTV application, security managers should keep in mind the following points:

- CCTV is a visual tool of security and should be applied accordingly.

- The application dictates the equipment, not the other way around.

- No matter what, the system will become obsolete.

- If a system is obsolete but still performing well, it is because the original application was correct.

---

[1] Information contained in this part of the *Protection of Assets Manual* is either a direct quote or an interpreted explanation of information from *The Professional Guide to CCTV* by Charlie R. Pierce, printed and published by LeapFrog Training & Consulting, copyrighted by C. Richard A. Pierce, used with permission.

- CCTV systems should always be designed with future growth and application changes in mind.

2.2. THEORY OF VISUAL SECURITY

Video motion is an illusion, a sequence of pictures flashed in front of the eye at a rate that the brain perceives as movement. Blank spaces between images add smoothness to a scene and bolster the illusion of motion. In analog systems, the monitor paints individual lines (horizontal sweep lines) across the screen one at a time, from left to right, top to bottom. It also paints an equal or greater number of lines up and down (vertical sweep lines). Where the horizontal and vertical sweep lines meet, one finds a point or pixel of energy. The more pixels a monitor displays, the better the overall resolution will be. For the most part, vertical resolution is restricted by NTSC[2] or PAL[3] standards. Horizontal resolution, on the other hand is limited only by the camera imager, monitor, and bandwidth of the transmission and recording medium. Consequently, the most common measurement of the quality or detail in an analog image is horizontal resolution. The more lines of horizontal sweep on a screen, the better the detail in the video picture.

All analog CCTV monitors and cameras employ a two-to-one (2:1) interlace pattern. The monitor first paints the odd-numbered horizontal sweep lines of the image and then resweeps the screen with the even-numbered horizontal sweep lines. This process creates 60 fields (half pictures) of information per second in NTSC and 50 fields per second in PAL. Combining one odd field and one even field of video information produces one complete frame or picture of analog video. In NTSC, for example, the viewer thus sees 30 complete frames each second.

Digital technology does away with the 2:1 interlace. Digital images are presented on the monitor as a full grid of small squares. Digital video is not measured in terms of frames or NTSC or PAL standards. Around the world, digital standards vary, but digital technology can adapt quickly to changing needs.

2.3. REASONS FOR CCTV IN SECURITY

Security measures come in four broad categories:

- **Electronic security measures.** These include burglar and fire alarm systems for prevention, deterrence, and response.

---

[2] NTSC stands for National Television Standards Committee and is the standard for resolution in the United States, Japan, and parts of Latin America. The NTSC standard is 60 fields per second at 525 vertical lines.

[3] PAL stands for phase alternation line and is the standard for resolution in Europe, Australia, China, and parts of Latin America. The PAL standard is 50 fields per second at 625 vertical lines.

- **Access control systems.** These include all forms of locks and barriers, including doors, walls, floors, and ceilings, as well as electronic access control systems.

- **Professional response.** This might come from the police or private security officers, or it may be as simple as lights that turn on automatically in response to a predetermined situation.

- **CCTV systems.** These are the visual imaging systems that include the components discussed in this work.

CCTV systems are meant to be visual assessment or visual documentation tools—nothing more. Visual assessment refers to having visual information of an identifying or descriptive nature during an incident. Visual documentation refers to having visual information stored in a format that allows the study or review of images in a sequential fashion. In addition, visual documentation includes various embedded authenticity points, such as a time/date stamp or character generation.

The bottom line is that there are only three reasons for having cameras in security applications. Granted, there are hundreds of applications for cameras, but there are only three reasons:

- to obtain visual information about something that is happening
- to obtain visual information about something that has happened
- to deter undesirable activities

Satisfying the first two reasons requires the right combination of camera and lens. One should base the choice of camera first on the camera's sensitivity, second on its resolution, and third on its features. Sensitivity refers to the minimum amount of visible light that is necessary to produce a quality image. Resolution defines the image quality from a detail or reproduction perspective. The camera's features are the aspects that give one camera an advantage over another, such as video motion detection, dual scanning, and built-in character generation. The camera should be chosen before the lens.

It is common to use different camera models within the same system. However, one should not use cameras from different manufacturers within the same system. Even within the limits of the NTSC and PAL standards, each manufacturer produces an image slightly differently. Phasing and sequencing problems most often arise when cameras from several manufacturers are used in one system. Fortunately, phasing problems are diminishing due to the sophistication of controllers and the digital industry.

Lenses determine what amount and type of image will ultimately appear on the monitor. A lens should be chosen for its ability to produce the desired identification information. The three theoretical identification views of a CCTV system are subject identification, action identification, and scene identification.

Subject identification. This is the ability to identify something or someone within the scene beyond a shadow of a doubt. An illustration of subject identification would be to view a $100 bill from 3 feet away (0.9 m). A person with mediocre eyesight should be able to identify the item as a $100 bill without a doubt. This sort of identification, like CCTV identification, does not allow the viewer to touch, smell, or taste the object (and usually not to hear it), so the visual information must be sufficient for identification. If the $100 bill were moved to a distance of 30 feet (9 m), a person with normal eyes would not be able to identify it specifically.

Another demonstration of subject identification shows how the camera's angle of view affects the results available from a CCTV system. A person standing on a desk, looking directly down onto the top of various peers' heads as they come and go from an office, would not be able to identify the people—especially if the viewer did not know the people. Identification from such a steep angle is even more difficult with CCTV images. Thus, subject identification depends first on the size and detail of an image and second on the angle of view.

Finally, for subject identification, the object should occupy at least 20 percent of the scene's width. The average person is 2 feet wide (0.6 meters). Therefore, to show the full body and still make it possible to identify the person beyond a shadow of a doubt, the scene can be no more than 20 feet (6 meters) wide. This guideline is based on a minimum 325 horizontal line resolution, which is television quality. The designer must pay extra attention when using newer digital storage or projection systems. Quite often, the image on the screen has a considerably lower resolution. Although the image might appear sharp in its 1 or 2 in. square on a PC screen, it may become indistinct when enlarged to a 5 x 7 in. (12.7 cm by 17.8 cm) print.

Action identification. This form of identification captures what happened. For example, a person pins a $100 bill to a wall and steps back to watch it. A second person enters the room, and the first person closes his or her eyes. Now the $100 bill is gone. The first person saw the second person enter the room but did not see him or her take the money. Thus, the first person does not have enough evidence to prove that the second person took it.

The lesson is that CCTV systems should be automated through a trigger. A system can be programmed to respond to video motion detection, pressure on a floor mat, or the breaking of a photoelectric beam. The triggered response might be to improve the resolution or the number of images recorded per second. The response might also be to bring the image to the attention of a guard. With automated triggering, the system records important actions and captures useful evidence.

Scene identification. Each scene should stand on its own merit. If a security officer witnesses a fallen employee via CCTV, he might respond according to procedure and call paramedics. However, if the security officer does not read the character generation on the video screen, he or she might send the paramedics to the wrong location—that is, to a place that looks like, but is not, the location where the employee fell. Scene identification is an all-important but often missed form of identification.

The angle of view and the three forms of video identification dictate the placing of cameras and selection of lenses. In summary, the object of subject identification should take up at least 20 percent of the scene width. Action identification requires at least 10 ten percent of the overall scene. If both types of identification are needed, it may be necessary to use two cameras.

2.4. ANALOG SYSTEM COMPONENTS

An analog video system consists of three main components:

- camera (used to transform a reflected light image into an electronic signal)
- coaxial cable or other electronic signal carrier (used to transmit the electronic video signal from the camera to the monitor)
- monitor (used to translate the electronic video signal into an image on a screen)

Other parts of an analog video system include the following:

Pan or pan/tilt unit. If the user wants the camera to move, a pan or pan/tilt unit is in order. A pan unit moves the camera from side to side. A pan/tilt unit moves the camera from side to side and up and down. Pan/tilts can be an important part of a CCTV system, but in most applications it is more cost-effective to use several fixed cameras instead of a single pan/tilt camera. Today's pan/tilt systems, for the most part, are built into a single unit protected inside a dome (call an auto-dome). Originally, pan/tilts helped cut system costs by reducing the number of cameras needed. However, today it is possible to buy several fixed cameras for the cost of one pan/tilt system. Therefore, pan/tilts are now typically used mainly in systems that use zoom lenses, that interface with alarms,[4] or that employ pre-positioning.[5]

---

[4] In alarm interfacing, an event (such as tripping of a door contact, photo beam, or motion detection system) is used to trigger a response from the camera system. The response may be to call up an image to a specific screen or multiple images in a series or layout on the screen or to trip a recording system to a higher frame rate of recording. Alarm interfacing cuts down on the amount of times spent watching images.

[5] Pre-positioning is a mechanical or electronic setting that directs the camera to return to a particular pan/tilt and zoom position when a signal is tripped.

A system designer should not add a pan/tilt without carefully considering the application's demands, as pan/tilt systems may demand much staff time.

Controller. A controller commands a function of a pan unit, pan/tilt unit, or automatic lens. In selecting a controller, it is important to remember that the application chooses the equipment, not the other way around. Once the application is decided, the equipment will fall into place.

Switcher. To show the displays from several cameras on one or more monitors, a system generally requires a switcher. Video switchers save money by making it possible to use more cameras than monitors. They come in many types. With a passive, four-position switcher, the user pushes a button and an image appears on the screen. Dwell time is the time a sequential switcher automatically switches from camera to camera. It is not a factor when you have a switcher that has more inputs than cameras attached. A sequential switcher automatically switches from camera to camera. A quad splitter displays four images on a single screen. A user with a sequential or quad splitter should consider upgrading to a multiplexing switcher, which can interact with the system's video recorder to store more information per camera. A multiplexing switcher can also play back video streams separately or in quad format, according to need. Another type of switcher is the matrix switcher, which can organize large groups of video inputs and outputs and integrate them with alarms and viewing options.

Lens. If it focuses light onto a chip or tube within a camera, it is a lens. Lenses come in a variety of sizes, allowing many different fields of view (the width and height of the scene). The proper lens makes it possible to capture an image that provides the right amount of identification within the overall scene.

Video transmitter/receiver. This type of device allows the video signal to be transmitted via cable, phone line, radio waves, light waves, or other means. It is common to use several video transmission methods within a single system. Coaxial cable is the most common medium but not necessarily the best for all cameras within a system. It is possible to have a long, outside run on fiber-optic cable in the same system that uses two-wire (twisted pair) transmission as its primary transmission method.

Amplifier. Amplifiers strengthen the video signal for long-distance runs, but one should avoid amplifiers when possible, as most installers do not have the proper tools to balance amplifiers into the system. If a system uses a coaxial cable run that is long enough to justify the use of an amplifier, it may be better to replace the coaxial cable with fiber-optic, microwave, or two-wire transmission.

Video recorder. This device retains the video information on a magnetic tape, CD, DVD, hard drive, or other medium. The right form of recording is determined by the need. Recording features include time-lapse, event-triggered, 24-hour, 72-hour, and more.

Once the designer understands the basics of each preceding category, it is possible to design a simple or complex system on paper and have it work in the field.

The key points to remember are these:

- Once simplified, the most complex of electronic systems can be managed by almost anyone.
- The application drives the choice of equipment.

2.5. DIGITAL SYSTEM COMPONENTS

The world of analog CCTV is fading into the background behind digital. Even the name, CCTV, will change. A fully digital environment will use terms like digital imaging systems or visual imaging systems.

The three main parts of a digital video system are as follows:
- camera
- digital electronic signal carrier, such as Category 5 (Cat 5) cable
- PC with viewing or recording software (sometimes accessed via a Web browser)

Other parts of a digital video system include the following:

Digital electronic scanning software. These programs allow a fixed camera to appear as if it were mounted to a mechanical pan/tilt device. This is accomplished by scanning across the imager in a predetermined path as opposed to physically moving the camera. It is also possible to deploy digital cameras with traditional, mechanical pan/tilt devices inside a dome enclosure.

Controller. For the most part, digital controllers are computer programs installed on a personal computer, network, or handheld device. Digital controllers can work with a joystick or be controlled via pointing and clicking with a computer mouse. Most Internet protocol (IP) cameras[6] and digital systems use proprietary graphical user interfaces (GUIs).[7]

---

[6] Analog cameras are sometimes marketed as being "digital" if they have digital effects. Fully digital cameras may also be called Internet protocol (IP) cameras. IP is a language for digital transmission.

[7] A graphical user interface is the visible screen (like that presented by Microsoft Windows) used for controlling a computer.

Switcher. A digital CCTV system can use three types of switching. First, it might use a high-speed, analog-to-digital converter that accepts multiple analog signals and outputs a single, multiplexed[8] digital signal. Second, the system might use a high-speed digital switcher that doubles as a multiplexer. Third, some digital recorders contain built-in multiplexers. It is also possible for all camera outputs to be connected directly to a server. In that case, the PC is not so much a switcher as a controller and management tool. Once a camera is connected to a PC, the video information is stored on a hard drive, CD, or DVD. The information is not necessarily multiplexed but may be stored as individual sequences or image files.

Lens. The lens is one of the few elements of a video system that is not converting to digital. However, because of the capabilities of IP cameras and other control points within an IP system, various functions of the lens (e.g., auto-iris, zoom, and focus) can be automated so they do not require setup by a field technician and do not have any moving parts.

Video transmitter/receiver. Ethernet is the most common transmission medium for digital systems, though it is not necessarily the best application for all cameras within a system. It is possible to have a long, outside run on fiber-optic cable in the same system that uses two-wire transmission as its primary transmission method. It should be noted that RG-59/U, RG-6/U, and RG-11/U coaxial cables cannot be used to carry a digital video signal.

Amplifier. For the most part, amplifiers in the digital world are repeaters. Since digital signals are binary codes, the signal itself does not require amplification. It is the carrier[9] that must be maintained. A signal is received, and an exact copy with a renewed carrier strength is emitted.

Video recorder. Although digital video recorders (DVRs) made a huge entrance early in this millennium, many larger applications proved that stacked or multiple DVRs were not necessarily the way to go. DVRs have fixed inputs, accepting signals from 4, 6, 8, 12, 16, or more cameras. However, most DVRs are not true digital recorders, as they only accept analog inputs and their outputs are equally analog. In larger systems, DVRs are being replaced by massive hard drives with simple GUIs.

2.6. SYSTEM DESIGN

CCTV systems are not as complicated as they appear. System design can be addressed by following a few simple rules:

---

[8] Multiplexing combines the signals from several video cameras into a single data stream. The combined signal does not carry the full number of frames from each camera. If a system has three cameras, each producing 30 frames per second, a multiplexed signal of the three would carry 10 images per second from each.

[9] The carrier is the electronic signal on which the digital stream or sequence of electronic commands rides.

- **Keep the system in perspective.** CCTV systems contain only three major components: camera, cable, and monitor. Any other item is peripheral.

- **The application chooses the equipment, not the other way around.** Salespeople may claim their equipment will handle all the user's needs—without even asking what those needs are. The user should first determine his or her needs and only then select equipment.

- **Design generically.** The best system design does not specify models or brands but remains open to several bidders, who can offer to use whatever equipment they normally supply. An added advantage is that the designer, who might also be the user, does not have to be an expert regarding specific equipment models.

- **Anything can be done with the right resources.** It is best to design the application before establishing a budget. After the design is completed, the designer should calculate its cost and then, if necessary, remove some elements. One should always design for the best option first.

- **The system does not have to be built all at once.** Once the design is completed, the installation may be stretched over several budget years. The key is to work with equipment that is solid, is proven, and will most likely remain available for the next several years.

The user should take the following steps before putting the new system out for bid:

STEP 1: WRITE THE PURPOSE OF THE PROPOSED CCTV SYSTEM.

If the purpose is to monitor the back aisle of a store, little advance layout is necessary to make a viable system. If the intention is to cover several locations in several complexes, a bit more work is required. In any system, over time, parts of the system may end up serving purposes other than security. The designer should consider the many ways in which the system might be used. In the process, he or she may discover unknown options. For example, it might be possible to divide the system so that some segments serve security purposes and others monitor traffic or product flow. It may then be possible to split the CCTV budget with other company departments.

The bottom line is to write out the purpose. Having it written out will keep the designer on track and allow the upgrade to change in a logical way if necessary.

STEP 2: WRITE THE PURPOSE OF EACH CAMERA IN THE SYSTEM.

To define the purpose of each camera, it is necessary to weigh the security risk of each area to be viewed. For a high-security area, it may be worthwhile to interface the CCTV system with an alarm device, such as a door contact, microwave motion detector, photo beam, or video motion detector. In a low-security area, alarm interfacing may not be necessary and recording the video signal as a backup reference may be the best option.

Another issue is whether the unit should be visible or covert. Camera visibility can, to a limited degree, prevent certain nonchalant crimes or crimes of convenience in shopping centers or malls. Covert cameras, on the other hand, often promote security while allowing individuals to be comfortable in their surroundings. If covert applications are to be used, the security manager must make sure to respect privacy rights.

STEP 3: WRITE THE AREAS TO BE VIEWED BY EACH CAMERA.

In this step, one defines the proposed cameras' actual views in terms of both height and compass points. For example, the designer should write that camera 1 will be 30 ft. high on the northeast corner of building 5, looking at the west gate, and that camera 2 will be mounted above the west exit in the north hallway of building 6, looking east.

STEP 4: CHOOSE A CAMERA STYLE.

The choice of camera style should be based on sensitivity, resolution, features, and other design factors:

Sensitivity. Sensitivity refers to the minimum amount of light required by the camera to produce an image. The first consideration in choosing a camera, then, is the lighting in the area. Is it bright or dim? Is it constant or variable? Does the location contain lights that could brighten the scene at night or during cloudy days? Does the viewing area contain large windows? Are the windows covered with heavy curtains? Are the curtains closed during the day, or are they opened according to the varying light conditions outside? Will the proposed scene have a bright background that silhouettes the people being monitored?

For interior cameras, sensitivity is not usually a major concern unless, for example, the areas viewed by those cameras are not lighted adequately at night. If the system design includes several exterior cameras, a lighting study may be in order. Such a study is not especially difficult but requires some training and a good light meter. If additional lighting is necessary for nighttime operation, one can consider both visible and infrared (IR) lighting. IR light is not visible to the human eye, but many cameras view it the same way humans view visible light. If the decor of a building or the presence of neighbors forbids the installation of visible lighting, IR lighting may be the solution.

Cameras come in three basic sensitivities: full light, lower light, and low light. *Full-light cameras* are designed to produce good, even pictures indoors under full, consistent or minimally variable lighting conditions. *Lower-light cameras* are designed to produce good, usable video images under lighting conditions equivalent to dusk or dawn. *Low-light cameras* are designed to produce images where little or no light exists. Full-light cameras are usually the least expensive, lower-light cameras are in the middle to high price range, and low-light cameras are the most expensive.

Resolution. Resolution is a critical measure of the picture quality and specifically is the number of horizontal scan lines or digital pixel arrays that the camera captures. The cameras selected must capture video at a resolution sufficient to produce images with enough detail to produce viable visual evidence.

Features. CCTV cameras offer numerous special features that the system designer can consider. The following descriptions explain some of the most popular features:

- **Automatic gain control (AGC).** An AGC circuit is built into most cameras that have a wide range of sensitivity. This is an internal video amplifying system that works to maintain the video signal at a specific level as the amount of available light decreases. AGC was originally designed to ensure that a camera continued to produce a consistent image as it panned through a shaded area. All cameras mounted outside should have the AGC switched on. Many charge-coupled device (CCD) cameras sold today do not give an option of turning the AGC off.

  Unfortunately, AGC increases noise in the video picture by a factor of 10, degrading the video quality dramatically in low-light situations. However, AGC is extremely useful when the camera swings into an area that is just below minimum light requirements, allowing the security person to continue viewing in areas that normally would go black.

  A camera's AGC sensitivity should not be confused with its general sensitivity. AGC sensitivity, which provides a usable but low-quality image, is sometimes quoted as a way to make cameras appear, on paper, more sensitive than they really are.

- **Electronic shuttering.** Electronic shuttering (manual or automatic) refers to a camera's ability to compensate for light changes without the use of automatic or manual iris lenses. The feature is analogous to using eyeglasses that turn dark when exposed to bright light. Such glasses reduce the amount of light that reaches the eye, just as electronic shuttering reduces the amount of light that reaches the camera's imaging. With electronic shuttering, a camera can either work in a wide range of light without an auto-iris lens or produce high-speed images to capture fast-moving objects.

  With the introduction of the IP camera and the digital hybrid, electronic shuttering changed. No longer is the image darkened as a whole. Now each pixel on the CCD or imager is analyzed and filtered. If the image has a bright spot, the electronic shutter dims the points of extreme brightness without affecting the overall performance of the camera or reducing its sensitivity. This can all be done without an auto-iris lens. However, it is still advisable to field test the equipment to verify its claimed specifications.

- **Backlight compensation.** One of the hardest tasks for a CCTV camera is to view a subject in front of a bright background. The most common such situation involves looking at someone standing in front of a glass door to the outside. Manufacturers have developed various methods for meeting that challenge:

  — **Auto-iris lens.** The most common tool for controlling the brightness of an image focused onto a chip is the auto-iris lens. Unfortunately, many CCTV system designers misunderstand how an auto-iris lens works. This ignorance often leads to camera applications that are beyond the scope of the camera's ability, resulting in silhouetted or extremely dark images.

  All electronic, auto-iris lenses are designed to respond to the average amplitude of the raw video signal produced by a camera through a video sampling circuit. As a CCD (imager) is exposed to more light, the raw video signal increases. As the light decreases, the raw video signal decreases. The auto-iris lens, if installed properly, ensures that the video image remains at an average of one volt peak-to-peak (vpp) under optimal lighting conditions. As the video signal increases or decreases, the auto-iris lens closes or opens in direct proportion. For normal lighting conditions and fluctuations, this method of control works well. However, since the video sampler works on averages, the camera staring into bright areas will compensate for the brightness. This leaves all other portions of the image dark or in silhouette. That is why auto-iris lenses cannot keep up with this everyday problem application.

  — **Peak average (P/A).** P/A is method of auto-lens manipulation in analog hybrid cameras. It is a limited method of lens manipulation with good results in a few cases. P/A works by averaging the video signal, dark and bright, into one overall signal. If the dark areas are to be brought out, then the bright areas go into overload and wash out. If the bright areas are to be brought out, the dark areas go darker still and lose detail. When the signal is averaged and the decision to improve bright or dark is made, the iris of the lens is forced open (or closed) slightly. Opening it allows the front or dark portion of the image to have more light reflected into the camera. Closing it is favorable for the bright areas.

  A problem comes from the antiquated method of electronic averaging. P/A requires an image that is two-thirds dark area and one-third bright area or vice versa. It does not work on a 50/50 image. Thus, its applications are limited. P/A is declining as more sophisticated methods become available.

— **Masking.** This method of digital interfacing with the video signal is built into specific cameras and controllers. Masking divides the video image into grid sections. Next, various sections are programmed to be ignored. Then the grid overlay is turned off and the image is viewed without obstruction. Compared to P/A, masking is more effective and controllable, though it produces similar dark and bright areas.

— **Electronic iris.** This is the first true method of digital signal enhancement that obviates the need for auto-iris lenses. Unlike its predecessors, the electronic iris works on true video signal averaging. This form of electronic enhancement literally de-amplifies the super-brights and amplifies the sub-blacks, creating an equal, 1 vpp video image. The result is that a person standing before a bright glass door is fully visible in detail, as are the surrounding features of the image. Manufacturers use a variety of names for the process.

— **Super Dynamics.**[10] This method of electronic backlight compensation double-scans the CCD. The first scan is done at the standard 1/60th of a second, capturing images in the lower-light areas but washing out images in the bright areas. Simultaneously, the imager is scanned at 1/10,000 of a second, producing an image in the bright areas. Then the onboard processor combines the two images and the best of each is kept while the rest is discarded. The net result is a single image that has the best of both worlds. However, this technique does not work in all situations.

Cameras use various methods for handling difficult lighting situations, and manufacturers use different names for those methods. Few methods, however, are more effective than aiming the camera away from direct light. Eventually, obtaining good video in multiple lighting situations will be as simple as hanging a camera on the wall and aiming it correctly.

— **Auto focus.** This term refers to a camera's ability to focus on a scene automatically. While tracking an action or changing a scene, the camera and lens work in concert to focus the image. However, if the camera is pointed at something that is tucked into the middle of several objects that are closer, the electronics may have a problem determining what to focus on. Dome covers can also interfere with auto focus. For the most part, however, digital hybrids and auto focus IP cameras work well. The best equipment provides the option of turning auto focus off so the operator can focus manually in difficult situations.

— **Privacy blocking or image protection.** Through the addition of a specific control or a digital effect built into the camera, it is possible to obscure a specific area within an image. It only takes one liability case to realize that there are people behind the pan/tilts and zoom lenses. A few of the millions of CCTV cameras installed around

---

[10] Super Dynamics is a trade name for a particular Panasonic technology.

the world stray from their intended uses and become invasive toys. Even in public settings, minor invasions of privacy can escalate into major problems, usually after captured images are made public.

Privacy blocking makes it possible to block specific elements of a scene, such as particular windows. Thanks to digital technology, the mask that is overlaid onto the image is mathematically attached to the relevant pixels, so the system works even when the camera pans, tilts, or zooms. This powerful feature is affordable and comes with various options, such as locked positioning or the ability to pan/tilt or zoom in or out while continuing to mask the desired area; flexible drawing or the ability to mask a section of the image regardless of its size or shape; blackout or focus tampering of the specified area; password protection against unauthorized reprogramming of the protected areas; and the ability to temporarily remove the mask or view through it during playback (after the fact) for valid purposes under controlled circumstances.

- **Other design factors:**

  — **Environment.** Does the camera need to be protected from vandals, customers, or employees? Must the camera housing blend in with existing decor? Interior camera housings are available in configurations ranging from ceiling mounts (designed to mount in place of ceiling tiles) to overhead bubbles to corner-mount stainless steel or bronze housings. Is the environment extremely dirty? Excessive dirt or dust buildup in the camera causes the unit to run hot and may limit its performance or life span. If the camera is to be mounted outside, other considerations arise. If winter temperatures in the locale drop below 35° F (1.7° C), the camera's protective housing will need a heater. Although CCD cameras can operate in cold temperatures, lenses tend to slow down or freeze up because of the type of grease used in the iris, focus, and zoom rings. If summer temperatures rise above 80° F (26.7° C), the housing may need a fan.

  If the unit will be installed in the sun, a sun shield will be needed to prevent excessive heat buildup in the camera housing. If the unit will face east or west, a sun visor may be needed to cut the glare from the rising or setting sun and to deter overheating. Often a sun shield can also act as a visor.

  — **Mounting.** Mounting raises several considerations. If a camera will be mounted up high, how can it be reached for maintenance? Will the camera's angle of view be so steep that only the top or back of heads will be visible? Will the unit be mounted under an overhang, next to a wall, or above major obstacles such as air conditioning units? Will there be enough room to open the housing from the top, or will a bottom-hinged unit work better? Should the unit be mounted on a swing arm to allow easy access from a window on upper floors? If the unit is to be mounted in a ceiling tile housing, is there enough room above the false ceiling to install the housing?

Another mounting consideration concerns what each camera is required to view and from what distance. For proper security viewing, one should not depend on the camera to view more than two objectives (one major and one minor), and the camera should not auto-pan (move side to side), either physically or digitally, more than 45 degrees left or right from the center of its major focus. Good video surveillance is often waylaid by expecting too much from a single unit or installing fewer cameras than are actually needed. The objective of the video system should be remembered at all times. The higher the security risk of the viewing area, the fewer objects each camera should be required to watch. In high-security locations, it takes at least four cameras to view a 360° area, though the budget may force a compromise.

STEP 5: CHOOSE THE PROPER LENS FOR EACH CAMERA.

This choice is determined by three different factors: camera format, distance from the camera to the scene, and field of view.

Format size. Format refers to the size of the imager area onto which the lens focuses light. That size is measured diagonally. The format size of the lens must match or exceed the format size of the camera. If the lens's format size is too small, the lens will not fill the imager with a picture, and the result will resemble tunnel vision. Thus, a 1/4 in. format camera requires a 1/4 in. format lens or larger (such as 1/2 in., 2/3 in., etc.).

Distance from camera to scene. This factor determines what focal length of lens is needed. The distance is measured from the front of the camera to the main subject being viewed. This distance must be measured accurately. If the camera will be mounted on the side of a building, 40 ft. from the ground, and the center of the scene is 30 ft. from the building, the relevant distance is not 30 ft. but 50 ft. The Pythagorean theorem shows that A^2 (the square of the distance from the side of the building to the center of the scene, or 30 ft. x 30 ft.) plus B^2 (the square of the mounting height, or 40 ft. x 40 ft.) must equal the square of the distance from the camera to the scene. This example results in an equation of (30 x 30) + (40 x 40) = 2,500, which equals 50^2. Thus, the distance from the camera to the scene is 50 ft.

Field of view. The field of view—that is, the height or width of the area being viewed—determines the appropriate focal length for the lens. In the previous example, the distance from the camera to the center of the scene was 50 ft. Perhaps the desired scene is 20 ft. wide. With the use of simple math, a field-of-view calculator, a cheat sheet, or a view finder, it is possible to calculate the appropriate lens focal length. It is important to note, however, that the more area the camera views, the less detail it picks up.

STEP 6: DETERMINE THE BEST METHOD FOR TRANSMITTING THE VIDEO SIGNAL FROM THE CAMERA TO THE MONITOR.

This step may be difficult for a typical security manager, who might prefer to leave it to the bidding contractor. Coaxial cable is generally sufficient for analog cameras but does not work for IP-based systems. For distances of 1,000 ft. or more between the camera and the control point, it may be best to use fiber-optic cable, regardless of the type of camera. Many transmission methods are available, and each has its advantages, disadvantages, and costs. Among those methods are coaxial cable, fiber-optic cable, twisted pair (two-wire) cable, Category 5 (networking) cable, microwave technology, radio frequency or other wireless technology, infrared transmission, and transmission over telephone lines, the Internet, or an intranet. A system might use more than one method of video transmission.

STEP 7: LAY OUT THE CONTROL AREA AND DETERMINE WHAT ENHANCEMENTS ARE NEEDED BASED ON EACH VISUAL ASSESSMENT POINT'S REQUIREMENTS.

Step 2 defined the purpose of each camera. With those definitions, the system designer can assign triggers and priorities and then determine which features the control equipment must have. For example, the designer might conclude that camera 1 requires digital video motion detection, an expandable pole, and an alarm interface; camera 2 requires a pan/tilt and zoom controller with pre-positioning; camera 3 requires alarm activation via a door contact; and the manager's office in building 5 needs a slave controlling system with password access.

Determining triggers and priorities for each camera helps automate the video system. It is inefficient and ineffective to assign people to sit and watch tens or hundreds of scenes continuously.

A recording or storage system is also needed. Such systems are available in many formats. For the most part, standard VHS and SVHS tape systems have fallen by the wayside in preference to digital recording.

2.7. EQUIPMENT SELECTION

This section discusses CCTV equipment in greater detail. The point is not to choose the equipment vendor but to define the operational parameters required by the application.

CAMERA

CCTV cameras come in four main types. Understanding the distinctions makes it possible to select the right camera for the task and avoid spending more money than necessary by buying unneeded features.

Standard analog CCD cameras. These may be black-and-white or color. The most common type of camera, they work well in all indoor and many outdoor applications. They are analog-based and may or may not have digital effects. Resolution ranges from 220 horizontal lines (very low) to 580 horizontal lines (very high). Light sensitivity varies between .005 lux (.00046 foot-candles), which is very low, to 10 lux (.929 foot-candles), which is very high. Color cameras are the most restricted by low-light situations. To compensate for that limitation, manufacturers have developed hybrid analog cameras. Some use infrared sensitivity to capture more light. Others combine color and black-and-white capability in one unit, capturing color images during daylight hours and black-and-white images at night when the light is low. Other cameras use an intensifier between the lens and the CCD to amplify the available light tens of thousands of times.

IP cameras. These digital cameras come in black-and-white or color. Like their analog counterparts, IP cameras require visible light to create an image. These cameras are available in three basic styles: standard, megapixel, and smart. All IP cameras measure their resolution as a multiple of the common intermediate format (CIF), which is a resolution of 352 x 240. Standard IP cameras range from one-quarter CIF to four times CIF. Megapixel cameras range from 16 to 32 times CIF or higher. Smart cameras can fall under either resolution range. They are called smart because they take advantage of their server base and employ computer programs within the cameras. Those programs enable the camera to perform various functions, such as digital video motion detection, facial recognition, privacy blocking, digital pan/ tilt and zoom, and more.

IP cameras can typically transmit in analog as well as digital format. They may be powered via transformers or universal serial bus (USB) connections.

Infrared (IR) cameras. These cameras require an IR light source to create an image. They are used where visible light is not an option.

Thermal cameras. These require no visible or IR light to produce an image. Using special filters and lenses, the cameras monitor the temperatures of the objects in their field of view and use colors to represent temperatures. Cold objects are shown in varying shades of blue, while hot objects are shown in varying shades of red. Thermal cameras are often used in long-range surveillance, such as monitoring ships in a harbor five miles out. Since these cameras require no light to create an image, they are popular with police and border patrols.

LENS

After choosing a camera, choosing a lens is the second most important decision of the project. The selection depends mainly on the size of the scene and the degree of visual identification required. Lenses come in five main types: wide angle, standard, telephoto, varifocal, and zoom.

Wide-angle lens. A wide-angle lens captures a very wide scene and thus is best suited for short ranges—that is, 0 to 15 ft. (0 to 4.5 m).

Standard lens. For an average scene at medium distance, a standard lens is needed. This type of lens reproduces a view equivalent to what the human eye sees at the same distance, except that the human eye has peripheral vision approximately two and one-half times that of a standard lens. A medium distance would be considered about 15 ft. to 50 ft. (4.5 m to 15.25 m).

Telephoto lens. If the required view is a narrow area at long range, a telephoto lens is needed. A telephoto lens can best be compared to a telescope, as it enables the user to look at objects far away as if they are close. However, such magnification narrows the field of view. Long range is considered anything over 50 ft. (15.25 m).

Zoom lens. A zoom lens incorporates moving optics that produce the same views provided by wide-angle, standard, and telephoto lenses, all in one device. These lenses may be manual or motorized. With a manual zoom lens, the installer manually adjusts the lens's field of view and focus. Once set, the lens remains fixed. A motorized zoom lens has motors installed within the housing of the lens. This lens allows an operator, via a controller, to adjust the lens's optic view from a remote location. The motorized zoom lens also features a tracking mechanism, which is a physical tie between the focal optics and the zoom optics designed to automatically adjust the focus of the lens as the lens is zoomed out (from telephoto to wide angle).

Varifocal lens. This is a smaller version of a manual zoom lens, offering the opportunity to tune the view on-site. These lenses enable installers to carry a few lenses in stock to cover a multitude of scene ranges. Varifocal lenses differ from zoom lenses in two ways:

- They do not cover a full range from wide angle to telephoto but only a slight range on either side of a fixed focal-length standard lens (i.e., 8 mm to 12.5 mm).

- They do not have a tracking mechanism and must be refocused each time their range is changed.

The next factor in choosing a lens is compatibility between the lens and the camera. Not only have cameras become smaller, but technology has changed lenses' ability to pass light; ability to reproduce a detailed image; size; electronic controls (now moved from the lens to the camera); and cost. Thus, lenses must meet several criteria to match a camera's physical and electronic needs. Incompatibility can cause physical damage and image problems. Over time, compatibility problems are being designed out of the industry, but meanwhile the following questions must be answered to determine a lens's compatibility with the application and the camera:

CCTV

Question 1. Will the camera be installed in an area where the lighting is fixed, minimally variable, or highly variable?

This question determines whether the application requires a lens with a fixed iris, manual iris, or auto iris. As lighting has the greatest impact on performance, a fixed-iris or fixed-aperture lens has no adjustable physical control over the amount of light that passes through it. A manual-iris or manual-aperture lens can be opened or closed during installation to increase or decrease the amount of light that passes through it. An auto-iris lens uses a motor to open or close the iris. The need for more or less light is determined automatically by the video sampling circuit in the lens or camera.

Auto-iris lenses have become smaller, lighter, and less expensive than manual- or fixed-iris lenses. An auto-iris lens costs only a fraction of the price of a fixed- or manual-iris lens. In many applications, it is financially prudent to use auto-iris lenses on all cameras.

However, it may make sense to use a fixed- or manual-iris lens if the camera has a large light range, auto electronic shuttering, or an electronic iris. An auto-iris lens could conflict with the camera's electronics, causing extreme image flutter or total blackout.

Question 2. Is the camera a 1/4 in., 1/3 in., 1/2 in., 2/3 in., or 1 in. format?

The format size of the lens must equal or exceed the format size of the camera. However, the format size of the lens has nothing to do with the final size of the image, which is determined by the focal length of the lens.

Question 3. Is the camera a color camera?

If so, a color-corrected lens must be used. A color-corrected lens has been ground differently and has a special coating on the optics of the lens. These measures ensure that the focal points of all colors in the visible light spectrum come to the same image focal point. Many older black-and-white cameras use non-corrected lenses. Therefore, reusing lenses may create problems.

Question 4. Is the camera a C or CS standard camera?

CS cameras were brought into production during the late 1990s. Their CCDs were moved 15 mm closer to the lens mount, so CS lenses do not penetrate as deeply into the camera. However, many cameras meeting the older C standard are still in use, and lenses meeting the C standard are still being sold. Thus, it is important to check whether the camera accepts C or CS lenses. Mounting a C lens onto a CS camera without an adapter ring will crack the CCD. However, with an adapter ring, the image and camera will be fine. Mounting a CS lens on a C camera causes incurable focus problems. Some cameras can work with both C and CS lenses.

Question 5.   Will the camera accept an AC/EC (video) or DC/LC lens?

AC stands for auto circuit, and EC (electronic circuit) is its metric equivalent. DC stands for direct circuit, and LC (logic control) is its metric equivalent. For the most part, the DC/LC lens has become the standard for camera and lens design.

The AC/EC (also called video) lens uses the original, antiquated auto-iris lens design. It has a video sampler circuit built into its body to control the iris based on the video signal level of the camera. To cut the cost and increase the efficiency of auto-iris lenses, the DC/LC was developed. This lens has no electronics built into it for iris control but rather depends on the camera to supply it with the necessary positive or negative motor voltages to operate the iris.

Older cameras accept and operate video lenses but not DC/LC lenses. Most cameras designed after 1995 and before 2002 accept either the video or DC/LC lens. Cameras designed since 2002 accept only the DC/LC lens.

Question 6.   Does the application use infrared (IR) enhancement lighting?

If so, the camera should use an IR-corrected lens. Like a color-corrected lens, an IR-corrected lens uses special optics and coatings to ensure that the longer IR light waves are focused on the CCD.

In the end, compatibility problems between cameras and lenses can be avoided through careful attention to camera and lens specification sheets. Camera specification sheets tell which lenses the camera will accept. Lens specification sheets tell whether the lens will live up to the demands of the camera. It is not essential to understand all the processes that make a piece of equipment work. However, it is essential to understand the basic principals and nomenclature.

2.8. CAMERA FORMATS AND LENSES

As chips shrink, it becomes increasingly difficult to match lenses with cameras. For instance, a site might have 1/4 in. cameras with 6.5 mm–65mm zoom lenses. If the security manager wants to add a 1/3 in. camera to the site, he or she must determine what lens on the 1/3 in. camera will provide the same view. The key is to know what the standard lens for a camera is and then calculate from there. A 25 mm lens is the standard lens for a 1 in. camera.

(Again, a standard lens recreates an image equivalent to what the human eye sees at the same distance.)

A 1/2 in. format CCD is half the size of a 1 in. format CCD. Therefore, the standard lens for a 1/2 in. camera should be 12.5 mm—half the size of the standard lens for a 1 in. camera. Likewise, the standard lens for a 1/3 in. camera is one-third of the 1 in. standard, or 8.333, rounded to 8mm.

It is a little more difficult to determine the right lens if a 1/2 in. camera in the field, using an 8 mm lens, will be upgraded to a 1/3 in. format, and the same image size is desired. The first step is to convert the 1/2 in. camera and lens into equivalent terms for a 1 in. format camera and lens. A 1 in. camera is twice the size of the 1/2 in. camera, so the 1 in. camera's lens would be twice the size, too. The 8 mm lens thus equates to a 16 mm lens on a 1 in. camera. The next step is to convert the 1 in. equivalents into a 1/3 in. format equivalent. Thus, 16 mm divided by three equals 5.33 mm. In this case, the best choice might be a 1/3 in. varifocal lens with a range that includes 5.33 mm. An alternative would be to select a lens with a fixed focal length close to 5.33 mm, in a 1/3 in. format or larger, of course.

The field of view is the final size of the viewing area as measured in width and height. Analog systems create rectangular images that are four parts wide by three parts high. If an image is 12 ft. wide, it must be 9 ft. high. The field of view for IP or digital cameras works in much the same manner except that the image is no longer fixed at 4 x 3.

Resolution of digital cameras is measured in terms of the common intermediate format (CIF). The resolution ratings of all digital cameras are multiples or divisions of CIF. The most common IP resolutions (to date) are as follows:

Ratio	Label	Horizontal x Vertical Array	Grid
CIF	CIF	352 x 240 pixels	84,480 pixels
Quarters of CIF	Q CIF	176 x 120 pixels	21,120 pixels
4 times CIF	4 CIF	704 x 480 pixels	337,920 pixels
16 times CIF	16 CIF	1,280 x 1,024 pixels	1.31 megapixels
25 times CIF	25 CIF	1,700 x 1,200 pixels	2.04 megapixels
36 times CIF	36 CIF	2,112 x 1,440 pixels	3.04 megapixels

A 4 megapixel image (2,000 x 2,000) has the same resolution as 400 ASA film. A 6 megapixel digital image has the resolution of 100 ASA film.

The resolution of an image is determined first by the camera, second by the transmission method, third by the weakest link in the video system interface, and fourth by the reproduction capability of the image storage system. The higher the resolution, the sharper the image.

High-resolution cameras produce low-resolution images if low-resolution monitors are used. The monitor, however, is seldom the problem. Analog video recorders average a playback of 325 horizontal lines. Thus, the camera's high-resolution image may look good on the monitor but poor when a recording is played back. Multiplexers are another source of loss. These units take the video signal in, digitize it for features and switching, then turn it back to analog, losing up to 25 percent of the resolution in the process. Moreover, coaxial cable can cost another 10 percent to 15 percent of resolution due to sloppy installation, bad connections, and cheap cable.

Digital resolution, by contrast, is ultimately unlimited and extremely flexible. A digital image can be made smaller in size but with a greater number of pixels per inch. The effect is to reduce the necessary storage space to keep the image and the required bandwidth to transmit it—without loss of detail. However, it is not possible to take a low-pixel image and add pixels to improve it.

Resolution is not a major issue in most indoor applications and many outdoor applications. Still, if a security manager will rely on recorded images for information, he or she should ensure that the object of interest is large enough to be identified clearly. A high-resolution camera is most likely to be needed for scenes viewed through telephoto lenses.

2.9. SWITCHING SYSTEMS

A switching system is the first point of automation within a CCTV application. It controls what the user is able to see, record, and respond to. In choosing a video switcher, it is important to remember, again, that the application dictates equipment selection, not the other way around. The following questions can be used to find the perfect switcher for the job:

- **How many cameras will be switched at each monitor point?** Most video switchers come with an even number of camera inputs. If a new system has three cameras, a four-input switcher will be needed (unless the security manager wants to place a special order). If the system might expand to six cameras within the next two years, it would make sense to buy a six- or eight-position switcher.

- **Will any of the cameras be viewed at more than one monitor point? If so, will the individual monitor points need separate controlling capabilities?**[11] The purpose of

---

[11] Controlling capabilities include the ability to pull up individual pictures or control lenses or pan/tilt units.

this question is to determine whether a secondary monitor point should be handled by a looping[12] or bridging[13] switcher. If several monitor points are needed, a single switcher may be ordered as a looping, bridging switcher, which passes the video from a camera directly through the primary switching point to a secondary point, while providing one switching and one fixed monitor point at the primary monitoring point.

- **Will there be any interfaced alarm trigger points?** As a video system grows, it may become advantageous to install alarming devices (such as door contacts, infrared motion detectors, and photo beams) to trigger a single camera onto the screen when an intruder or motion is detected in an area. An advantage is that the operator can concentrate on other tasks until the alarm is triggered.

- **Will several monitors be connected to the same switcher?** If the user will need four, five, eight, or 68 monitors tied into the same switching point with individual control for each monitor as to the cameras viewed, dwell time,[14] alarm input, and other factors, it would be best to consider microprocessor-controlled video switching systems.

- **Will it be necessary to have individual switching or dwell times for each camera?** Most switchers provide a single, constant switching time for each camera. If the video system under design requires each camera to have its own switching time according to priority, it will be important to note this when ordering the equipment.

- **Will the switcher be required to trip any other devices, such as recorders, buzzers, or lights, in the event of an alarm?** For example, after a camera's image is pulled up on the screen, a switcher can signal a video recorder to switch from time-lapse to real-time speed, enhancing the recording with added information.

- **Are there plans to expand the system in the future? If so, can the switching system handle the expansion?** If the system being designed has four cameras, a four-position switcher may be all that is needed. If, however, the system will be updated to eight cameras within a year, it would be more cost-effective to buy an eight-position switcher up front.

- **Will the unit need to be rack-mounted or placed on a table top?** The answer influences some ancillary decisions in the selection of a switcher.

The final choice of a video switcher should not be made until the majority of the video system is already in place on paper.

---

[12] A looping switcher enables the user to pass video through the first video switcher to a secondary point without affecting the video signal.

[13] A bridging switcher has two or more independent outputs. The secondary output is typically locked onto a single camera at all times and is controlled by the operator at the switcher location.

[14] Dwell time is the amount of time that an image stays active on a screen or output. Individual dwell times can be set for each camera input on a switching system provided that the feature is available. Otherwise, the dwell time is equal for all cameras.

TYPES OF SWITCHERS

Switchers are defined by their size, their features, and whether they are analog or digital. The types are as follows:

Sequential (analog). This simple type of switcher may offer such features as independent dwell times; alarm input (dry contact[15]); alarm output (dry contact); skip[16]; hold[17]; looping, bridging, and looping/bridging; multiple camera; and cascading.[18]

Quad splitter (analog). This switching system allows four camera images to be displayed on a single screen simultaneously. It is usually used for simple applications and works on the same principles as a sequential switcher. A quad splitter with simple features may include alarm input (dry contact); alarm output (dry contact); looping, bridging, and looping/bridging; four or eight camera inputs; sequential switching through independent cameras; quad mode; and skip/hold in sequencing mode.

Multiplexer (analog). This high-speed, sequential switching system is designed to work with a video recorder. The dwell time per camera may be set according to how many images per camera are to be recorded or according to the recording speed of the recorder. The multiplexer is an advanced unit that allows image playback singly, in sequence, in quad, or in any configuration up to the number of camera inputs. Images recorded via a multiplexer must be played back through the multiplexer or a compatible demultiplexing unit. Multiplexers include such features as alarm input (dry contact); alarm output (dry contact); skip and hold without affecting the recording sequence; independent alarm response programming per camera (such as setting the number of images recorded per camera under alarm conditions); looping, bridging, and looping/bridging; numerous camera inputs; sequential switching through independent cameras; quad mode; and multiple images on a single screen up to the total number of inputs.

Multiplexers come in three basic formats:

- **Simplex.** This type of unit is designed to perform only one function at a time. With a simplex unit, one can record video or play back a prerecorded tape.

---

[15] A dry contact input is a switch that closes or opens without any voltage being applied to the contact points. It is equivalent to a light switch.

[16] Skip is a feature that allows individual cameras to be dropped out of the sequencing cycle while remaining available for alarm input response.

[17] Hold is a feature that allows the operator to lock onto a single camera input. Note: with simple switching systems, if a camera is placed on hold, it may be the only unit that is being recorded.

[18] Cascading is the ability to tie two or more units together so that more camera inputs can be accommodated.

- **Duplex.** A duplex unit is designed to perform two functions at once. For example, the user can record video on one recorder while playing back prerecorded tapes on a second recorder and monitor.

- **Triplex.** With a triplex unit, the user can enjoy the functions of a duplex unit and also monitor any camera or sequence of cameras on a separate monitor at the same time.

Multiplexer (digital). Digital multiplexers are not so much multiplexing units as switching systems. Their controllability is limited by the IP system they are connected to, and they come in two basic formats. The first format is an analog-to-digital conversion unit, which accepts analog inputs and converts them into a multiplexed digital signal. The second format is a digital-to-digital switch, which accepts IP inputs and multiplexes them together for transmission.

Many IP transmission formats are restricted to distances under 300 feet. A field multiplexer allows a central point to be used to combined several long runs into a single signal and in effect allow for doubling the distance with only a percentage of the cable. It is primarily used for combining multiple camera inputs and sending them to the head-end.[19]

Matrix switching systems (analog/digital). A matrix system can accept large numbers of camera inputs with large numbers of monitor outputs. These usually come in an 8:4 configuration (eight cameras to four monitors). However, once the initial configuration is set, any group of camera inputs or monitor outputs can be achieved by adding input/output cards. Matrix units usually come in three parts: central processing unit, camera input rack, and monitor output rack. Each camera input and monitor output can be fully configured as a separate, independent function. It is not uncommon to have multiple monitor outputs connected to various simple switching or multiplexing units. With analog systems, the controller is usually a board that allows for individually choosing a monitor output or camera input on the work screen. Digital units work via a mouse and computer screen.

2.10. RECORDING SYSTEMS

When it comes to retaining and using images of security events, the user must decide whether the system's purpose is to verify information, prove it, or aid a prosecution with it. This decision leads to the type of video imaging that is best for the situation. For example, if the video information is to be used in the courtroom, its admissibility may be determined by the quality of the recorded information, the way it was obtained, and proof of originality.

---

[19] The system's head-end is the point where all signals come together to be managed or manipulated. The head-end is where the controlling equipment can be found.

The following are the basic types of recorders:

Time-lapse recorders (analog). These recorders are designed to make a two-hour cassette record up to 900 hours by allowing time to lapse between recorded images. The chosen duration dictates how much information is recorded. Instead of a full 25 frames (PAL) or 30 frames (NTSC) of video information being recorded each second, a time-lapse recorder may capture only a fraction as many frames. The strongest market for the time-lapse machine is retail, industrial, and long-term surveillance.

Event recorders (analog). Event recorders are designed to record triggered events. They usually cost half to one-third as much as time-lapse recorders. They remain in standby mode, waiting for an event to record. When a dry contact closure comes to the unit's alarm input, recording begins almost immediately—there is no loss of the first three or four seconds of the security breach. However, since the number and duration of events recorded determines how much videotape is used, the recorder may run out of tape if it is not closely monitored. These units are most popular for covert surveillance, entrance monitoring, and other applications where a particular event is the desired subject.

SVHS recorders (analog). In the past, resolution loss during playback was a significant problem, and SVHS (Super VHS) was originally developed to reduce that loss in color recordings. Today, however, most high-quality, industrial-grade security recorders play back at a satisfactory resolution.

To create a color image, a camera produces two separate signals, luminance and chrominance. Luminance is the black-and-white portion of the signal, and chrominance is the color part of the signal. SVHS records each of these signals separately to avoid signal loss during playback. However, these machines are often falsely touted as being superior to standard recorders. In black-and-white recording, SVHS has no advantage over most high-quality standard machines. Further, most color security cameras have only a single video output, meaning the luminance and chrominance signals are combined in the camera. If the recorder splits them apart to record them separately, the degree of signal loss is the same as if a standard recorder was used. Finally, cameras recorded on an SVHS machine must be fully synchronized to the recorder. The only way to realize the advantages of SVHS is to use cameras that have a dual video output (of luminance and chrominance). Otherwise, these units have the same features, advantages, and disadvantages as regular industrial video recorders and a slightly higher cost to purchase and operate.

24-hour/72-hour high-density recorders (analog). These units capture a larger number of recorded images over a 24- or 72-hour period than do time-lapse machines. By changing the angle of the recording head and reducing the space between recorded images, the units capture three times as much information on an inch of video tape. However, to achieve 24 hours of continuous recording, the machines must use longer, thinner tapes (the T160 type). In practice, such tapes work poorly and users must resort to standard T120 tapes, which can record only 18 hours of video.

Digital video recorders (DVRs) (digital). DVRs capture digital video signals, not analog (unless the analog signal is first converted to digital format and compressed[20]). These recorders store video data on a hard drive, CD, DVD, or other medium. The challenge is that the video data requires a great deal of storage space. Therefore, DVRs compress the video image, using a particular codec (a compression engine or command sequence that causes the unit to combine colors, drop resolution, or both). Once compressed, however, the image quality may be poor. It is important to test DVRs before purchase. A more popular means of compression is to record fewer images per second. If the application is watching a dealer with a deck of cards, the DVR should record 30 or more images per second. If the application is watching people walk across a lobby, two or three images per second may be sufficient. Most DVRs can be programmed to record a different number of images per second from each camera input.

2.11. WHERE CCTV IS HEADING

The explosion in closed-circuit television will only continue to spread, both outward and upward, for the next 10 or 20 years. Security professionals' needs will help drive the future of CCTV. As applications expand, new equipment will be manufactured to meet those demands.

---

[20] Video compression takes a large amount of binary information and shrinks it by reducing the number of colors, lowering the resolution, or recording fewer images per second.

PART 3
High-Rise Structures:
Life Safety and Security Considerations

This text is excerpted from the February 2006 revision, published as "Part II: High-Rise Structures: Life Safety and Security Considerations" in the "Applications" chapter of the *Protection of Assets (POA) Manual*.

3.1. INTRODUCTION

This part applies the information contained in the chapters on physical security to address life safety and security considerations for high-rise structures. Additional material regarding the concepts and principles that are discussed can be found in other chapters of the POA.

3.1.1. WHAT IS A HIGH-RISE STRUCTURE?

Generally, a high-rise structure is considered to be one that extends higher than the maximum reach of available fire-fighting equipment. In absolute numbers, this has been set variously between 75 feet[1] (23 meters) and 100 feet (30 meters), or approximately 7–10 stories (depending on the slab-to-slab height between floors). As one source notes:

> The exact height above which a particular building is deemed a high-rise is specified by the fire and building codes[2] in the area in which the building is located. When the building exceeds the

---

[1] "NFPA 101 [*Life Safety Code*] defines a high-rise building as a building more than 75 ft (22.5 m) in height where the building height is measured from the lowest level of fire department vehicle access to the floor of the highest occupiable story. This definition is consistent with many model building codes, but it should be noted that many different definitions exist in local jurisdictions that use varying height and measurement criteria" (Holmes, 2003, p. 13-19).

[2] A code is defined as "a systematic collection...a private or official compilation of all permanent laws in force consolidated and classified according to subject matter" (*Black's Law Dictionary*, 1990, p. 257).

specified height, then fire, an ever-present danger in such facilities, must be fought by fire personnel from inside the building rather than from outside using fire hoses and ladders. (Craighead, 2003, p. 1)

High-rise structures first appeared in the United States around the end of the 19th century. Since that time their design and type of construction has changed, resulting in the skyscrapers that dominate the skylines of most major cities throughout the world.

Although the world contains types of tall structures, this chapter considers the high-rise to be essentially an office building, located in an urban or metropolitan area, accessible directly from one or more public streets, and having a mixed occupancy with public assembly areas, retail spaces, and conventional office use.

3.2. LIFE SAFETY CONSIDERATIONS

3.2.1. SPECIAL CONCERNS OF HIGH-RISE STRUCTURES

The risk, or "the possibility of loss resulting from a threat, security incident, or event" *(General Security Risk Assessment Guideline,* 2004, p. 5), is a function of factors such as the facility size, the number of occupants, and, for intentional threats,[3] the value[4] of the target.

For a high-rise structure the risks and potential losses are higher due to the large size of occupied floor areas, potential difficulties in containing and responding to threats, and the inability to effect immediate evacuation of an entire building. (Because of the limited capacity of building stairwells and elevators, not all occupants can leave a facility immediately. Also, in some incidents—especially fires and fire alarms—elevators serving the affected floors are not considered a safe means of occupant escape.)

The ability to mitigate threats for high-rise structures depends on structural design and the use of technology to detect and deter a threat, to communicate its nature and location, and to initiate automatic or organizational responses. The concepts for life safety protection are similar to those for the development of a security program: planning must address identified assets at risk, and solutions should follow the principles of deterrence, denial (prevention), detection, delay, and apprehension (removal of the threat). Life safety protection is primarily mandated by codes, particularly those for fire and building. At first glance this would appear

---

[3] "[A]n intent of damage or injury; an indication of something impending" (General Security Risk Assessment Guideline, 2004, p. 5).

[4] In the case of terrorism, the value of the target to terrorists may include its status within a city or country and its type of tenants. For example, if the targeted building is symbol of economic strength or is the tallest building in a city, a "successful" terrorist act will achieve on e the perpetrator's goals, i.e., widespread publicity.

to simplify decision making, but the standards[5] are designed for minimum protection only and their interpretation often requires a code expert and clarifications from the local authority that enforces the law or regulation.

Life safety issues have obvious impacts on security. The security manager for a building may be tasked with both functions even though, at times, the functions may have opposite goals. For example, life safety mandates that all occupants be permitted unimpeded egress from an area; however, security would rather not permit a thief, who has stolen property, to have such freedom to escape. Creative access control measures, conforming to applicable life safety codes, are often required. (See further discussion later in this chapter.)

Life Safety Aspect

The most critical threats[6] in high-rise structures include fire, explosion, and contamination of life-support systems such as air and potable water supplies. These threats can be actualized accidentally or intentionally, and because they propagate rapidly, they can quickly develop to catastrophic levels.

The most significant factors affecting life safety in high-rise structures are as follows:

1. Early detection and precise location of incipient hazards
2. Reliable communications throughout the structure and with the outside
3. Assurance of safe escape routes
4. Prompt application of appropriate control measures (such as fire extinguishment, containment or replacement of contaminated air, shutoff or filtration of drinking water, and containment and removal of explosives)

Fire is given special attention in high-rise facilities. Buildings tend to contain large amounts of combustible material, and the potential that a serious fire could move upward (particularly if the structure is not protected with an automatic sprinkler system) is ever-present. As one source observes:

> Despite the fact that fires are rare occurrences, if one does occur, everyone in a building must react quickly. In other emergencies, such as a winter storm or civil disturbance, the initial reaction to early warnings of this type of emergency will not necessarily determine its impact on the building. In a fire emergency, however, the first 3 to 4 minutes are critical. The timely handling

---

[5] A standard is "a model, type, or gauge used to establish or verify what is commonly regarded as acceptable or correct" (Webster's New Universal Unabridged Dictionary, 2nd ed., as quoted in the Protection of Assets Manual, 1999).

[6] Ensuring the safety of individuals can involve identifying threats that can cause physical or psychological injury. These threats may include crimes, such as robberies or assaults, or may be related to building emergencies, such as elevator entrapments, power outages, or natural disasters.

of a fire emergency according to sound procedures can help stop the event from rapidly becoming a major problem. (Craighead, 2003, p. 308)

Building Vulnerability Aspect

Structural integrity and essential utilities and services such as lighting, communications, elevators, and escalators can be threatened in the event of a fire or other incident. Various threats to high-rise structures can spread upward, downward, or horizontally. Fire, for example, can spread to higher floors because of the natural tendency of heated gases to rise. But if combustible materials (such as wall and floor coverings) follow a downward path, fire may spread in that direction. Fire may also spread horizontally within a floor, particularly in the absence of adequate floor-to-ceiling walls and partitions. In major conflagrations, without early containment and suppression, fire may spread in all three directions. With explosions or sustained conflagrations, partial or total building collapse may cause multiple injuries and deaths.[7]

Contaminants in the air supply, such as smoke[8] or chemical or biological agents, can travel with the air from the point of entry through its distribution pattern. If that pattern involves return and recirculation, contaminants can be repeatedly distributed over the same areas unless they are filtered or the distribution system is shut down. If a single air-handling unit serves multiple floors, contamination can affect all the floors served by that unit. Depending on the air handling system, contaminants may travel upward, downward, or horizontally on any involved floor.

Contaminants in the water supply move with the water from the point of introduction. Successful attack on a water supply system can be difficult because any opening in a charged pipe will cause a leak. But if a valve or control point is found and water flow is stopped temporarily, the contaminant may be introduced downstream of the valve and will move in the stream when the valve is reopened. If such a point could be found near the beginning of the distribution piping, it would be possible to contaminate a major portion of the structure. Gases and solids dissolved in a solution can be delivered in water to such outlets as sinks, drinking fountains, and toilets. Heavier contaminants might collect in traps and adhere to pipes but also might be released at outlet fixtures. For highly toxic agents, that degree of exposure could be enough to cause serious harm. Even if only a small quantity of a highly lethal agent was introduced, its effect could be widespread because of sustained use of the water before the contamination was detected.

---

[7] As was the case in the terrorist attacks on the World Trade Center, New York, February 26, 1993, and September 11, 2001, and the Alfred P. Murrah Federal Building, Oklahoma City, April 19, 1995.

[8] "[T]he total airbourne effluent from heating or burning a material" (Clarke, 2003, p. 8-9).

Communications facilities within a high-rise are usually routed through common vertical risers for the entire height of a structure and distributed horizontally from junction or service ports on each floor. In wire systems of this type, a complete severance of the applicable cable could halt all service downstream of the cut-point and could also affect use upstream by interrupting the direct path. Disruption of communications both inside and outside a building may have dire consequences, particularly affecting emergency communications. For that reason, alarm, signal, and communications systems should be distributed so that localized points can operate independently even when communication with a central control panel or processor has been interrupted.

Threat of Bombing

Life safety planning must also consider bombs. The risk that an occupant, visitor, or use of a delivery service could introduce a small quantity of explosive material into a building is ever-present. The use of screening equipment, such as explosive detectors, X-ray machines, and metal detectors, can mitigate such a threat. Rigorous vehicle control procedures can reduce the possibility of a large quantity of explosive material being transported into an under-building or subterranean parking area. However, in a crowded urban environment, it is difficult to prevent the destruction that could be caused by a bomb-laden vehicle parked close to a building.

For new construction, one possibility is to design structures to withstand the detonation of a defined quantity of explosive without progressive failure of the structure. Maximizing the distance between a parked vehicle and the building is important, as is window design. Unfortunately, such solutions involve cost that needs to be justified by the probability[9] of such an occurrence. For existing buildings, the cost of these solutions may be prohibitive, although some improvement in the distance between parked vehicles and the building can be achieved through the use of heavy planters, bollards, and other such barriers. Also, security film can be applied to existing windows to reduce the possibility of flying shards of glass, which can cause injury and damage.

In assessing whether a particular facility is at risk of a bombing, the following questions should be considered:

1. Is the building a likely target (such as being a landmark that may attract attention)?
2. Are any tenants in the building a particular target for domestic or international terrorists?
3. Is the building an obvious easy or soft target due to poor security measures?

---

[9] "[T]he chance, or in some cases, the mathematical certainty that a given event will occur" (*General Security Risk Assessment Guideline*, 2004, p.4).

4. Are any nearby buildings a likely target? (And, therefore, could collateral damage be caused by an explosion at a nearby building?)

In most cases, the likelihood that a particular building will be a target is minimal. However, emergency management plans, including the evacuation preparedness of tenants, should be up-to-date and rehearsed so that if such a threat does materialize, the building is prepared to respond properly. Such preparation is particularly important for high-rise structures, compared to low-rise. A single-story building provides speedy emergency egress for occupants via multiple horizontal paths, whereas egress from an upper floor of a high-rise structure necessitates the same horizontal paths of travel, plus vertical descent via elevators (if available and safe to use) and building stairwells.[10] Stairwells may become congested if a total building evacuation is occurring. Also, descending many flights of stairs may be physically taxing for office workers with full mobility and virtually impossible for the disabled or physically challenged without the assistance of others or the use of mechanical evacuation devices such as chairs.

3.2.2. DEALING WITH THE LIFE SAFETY PROBLEM

The threat of fire and fire-related events obviously affects both people and property. Many countermeasures that aid life safety have a direct impact on the protection of assets. Anything that assists safe response to any type of emergency contributes to total asset protection, in particular, the life safety of building occupants. However, some emergencies that appear as threats to life safety may be diversions designed to permit unimpeded access to, or attack on, physical assets. A good illustration involves a structure in which the building emergency exit stairwells are locked from the inside of the stairwell (in order to prevent reentry to a floor) but are equipped with a fire detection system that automatically unlocks the doors when a fire alarm occurs. A knowledgeable intruder could cause an event, perhaps by activating a manual fire alarm station,[11] that would trigger the automatic response and thereby provide access to a selected location via the stairwell. During such an incident, area occupants would more likely be concerned with responding to the apparent emergency and not with securing physical assets, making the attack more likely to succeed.

Such an illustration does not suggest that the fire life safety system should not be required to automatically unlock stairwell doors in such an emergency. The primary concern is life safety, and unless an emergency is clearly shown to be a hoax, life safety measures must take priority. The illustration does suggest that complete reliance on locked stairwell doors as an access control measure in a building equipped with automatic unlocking is not wise. Additional

---

[10] In many emergencies, particularly those involving the threat of fire, elevators are not considered a safe means of evacuation and therefore emergency exit stairwells are often the only means of escape.

[11] Manual fire alarm stations are also known as manual fire alarm boxes, manual pull stations, or manual pull alarms.

protection must be added to increase the reliability of the locked fire stairs[12] as a control measure. This problem is further discussed later in this chapter.

Local Code Compliance

The first step in assuring life safety is to comply with local building and fire codes applicable to the structure. However, a building, particularly an older structure, although conforming to local codes at the time of construction, later may not necessarily reflect either the most advanced or the best approach to life safety.[13] The building may require subsequent changes to bring it up to date and into current compliance.

In regard to cost considerations, meeting or exceeding current code requirements may achieve concessions elsewhere that offset the increased cost. For example, reductions in commercial insurance premiums or the stated amounts in funded reserves for the expense of retained risks may be available.

Fire Detection

This must rank as the first priority. Fire detection—more properly, the detection of products of combustion or the combustion process—is the first step in any response that ranges from confinement to extinguishment to evacuation and escape.

Fire detection must take into consideration the special nature of a high-rise structure; that is, it must be accurate in indicating where a fire is and what stage it has reached. Fire codes have long required that manual fire alarm devices be located on each floor of a multi-floor structure in the normal path of exit from an area. However, reliance on such a system alone is not sufficient. If, for example, someone who detects a fire on the fourth level of a facility immediately runs down the stairs and on the way out of the building activates a main lobby fire alarm pull station, what does that say about the fire? The fire alarm system could say "Fire in Lobby"—the logical inference from the location of the pull station. That might prompt the wrong response. A manual pull station ensures that people have an opportunity to signal a fire emergency and warn other occupants while escaping, but it does not ensure that the emergency area will be quickly identified. The answer does not lie in eliminating or relocating manual devices, but in augmenting the system with detection mechanisms that help determine the exact location of the fire emergency. The following means of fire detection should be distributed throughout a high-rise structure to optimize early and localized detection. (Please

---

[12] Fire stairs are "building emergency exit stairwells" (or more commonly "building stairwells") or a "stair tower." "The name reflects the traditional reason for emergency, which has been building fire" (*Emergency Evacuation Elevator Systems Guideline*, 2004, p. 45).

[13] In addition, municipalities, which adopt specific editions of standards into law, may not update the law in a timely fashion when the relevant standard changes.

see another chapter in the *Protection of Assets Manual* for additional information on fire detection sensors.)

Smoke Detectors. Smoke detectors are automatic fire detection devices designed to detect the presence of smoke. One source describes them as follows:

> Smoke detectors generally are located in open areas, spaces above suspended ceilings, spaces under raised floors (particularly in computer rooms and data centers), cafeteria areas, air duct systems, passenger and service/freight elevator lobbies, elevator shafts, elevator machine rooms, enclosed stairways, dumbwaiter[14] shafts, chutes, and electrical and mechanical equipment rooms. The specific locations and spacing of smoke detectors are determined by an assessment of local laws, codes, and standards and engineering issues. (Craighead, 2003, p. 187)

Even an incipient fire, controlled before structural burning, could have costly consequences in many interior spaces within a high-rise facility.

Heat Detectors. Heat detectors are automatic fire detection devices designed to sense a certain temperature or rapid change in temperature. The *Fire Protection Handbook*[15] notes:

> Heat detectors are very reliable and have the lowest false alarm rate of all automatic fire detectors. They are best suited for fire detection in small confined spaces where rapidly building high-heat-output fires are expected, in areas where ambient conditions would not allow the use of other fire detection devices, or where very early warning of fire is not required. (Roberts, 2003, p. 9-17)

Sprinkler and Standpipe Alarms. Where standpipe and hose systems and automatic sprinkler systems have been installed, water-flow and valve-position alarms are appropriate. A stand-pipe system in a high-rise building is designed to transport water vertically to upper floors of the building so that the fire can be fought manually with hoses. A sprinkler system is defined as

> a combination of underground and overhead piping that is connected to an automatic water supply and is installed throughout the building. The piping is specially sized or hydraulically designed with that portion of the piping within the building generally located overhead. Sprinklers are attached to the overhead piping in a systematic pattern and a valve controlling each riser is located either directly on the system riser or in the supply piping. The system is usually activated by heat from fire and discharges water over the fire area. A device actuating an alarm when the system operates is located on the system riser. (Puchovsky, 2003, p. 10-189)

Nature of Fire Alarm Systems. Modern fire alarm systems are computer-based and are similar in systems architecture to security alarm systems. Field devices, such as smoke detectors and air-handling controls, connect to field panels, which connect to the main computer. Some of the field devices may be connected in a long series but still be individually addressable, using, for example, individual names that they report when communicating their status.

---

[14] "[A] hoisting and lowering mechanism, used exclusively for carrying materials, with a limited size car that moves in guides in a substantially vertical direction" (Donoghue, 2003, p. 12-204).

[15] This chapter's quotes from the *Fire Protection Handbook* are used with permission from the National Fire Protection Association.

Some codes prohibit the fire alarm system from performing any function other than fire life safety and further prohibit the use of any component or device that has not been approved for fire life safety applications. Depending on local authorities, this may preclude the use of fire alarm systems to monitor or control security devices and may also preclude the use of a security system to monitor or control a fire life safety device.

Where such restrictions are in effect, the mandatory release of certain electrified locking devices (normally controlled by an access control/alarm monitoring system) during a fire emergency requires careful design coordination of the two systems. Typically, the fire alarm system may not send a signal directly to the security system or its field panels but will control an approved relay that connects to an approved lock power supply. All the associated wiring must meet code, and the lock itself must be fail-safe (i.e., designed to unlock under any failure condition, such as when power is removed).

Fire Extinguishment

Automatic sprinkler systems are an essential aspect of fire protection for high-rise structures. As the *Fire Protection Handbook* notes:[16]

> When sprinklers are present, the chances of dying in a fire and property loss per fire are cut by one- to two-thirds, compared to fires reported to fire departments where sprinklers are not present.... When sprinklers do not produce satisfactory results, the reasons usually involve one or more of the following: (1) partial, antiquated, poorly maintained, or inappropriate systems; (2) explosions or flash fires that overpower the system before it can react;[17] or (3) fires very close to people who can be killed before a system can react.

Other Fixed Systems. Standpipe and hose systems are typically found in tall and large-area buildings (Shapiro, 2003, p. 10-351). They consist of small hose connections and perhaps hose racks for internal use, along with larger connections for fire department use. The *Fire Protection Handbook* notes:

> Standpipe systems can significantly improve the efficiency of manual fire fighting operations by eliminating the need for long and cumbersome hose lays from fire apparatus to a fire. Even in buildings that are protected by automatic sprinklers, standpipe systems play an important role in building fire safety by serving as a backup for, and complement to, sprinklers. (Shapiro, 2003, p. 10-351)

In addition to water (the primary agent for extinguishing fires in high-rise structures), extinguishing agents include dry chemical and wet chemical systems, carbon dioxide, and halon and halon replacement systems,[18] which have specialized applications in such structures.

---

[16] "U.S. experience with sprinklers" by K. D. Rohr, National Fire Protection Association Fire Analysis and Research Division, as reported in Hall, 2003, p. 2-21.

[17] In the September 11, 2001, World Trade Center attack, the Twin Tower's automatic sprinkler systems, particularly those at or near where each plane hit, were destroyed by the impact or rendered ineffective by the burning fuel.

[18] Halon production ended January 1, 2000, "except to the extent necessary to satisfy essential uses, for which no adequate alternatives are available" (Taylor, 2003, p. 6-281). Existing stocks can be recycled until they are exhausted. A number of halon replacements have been developed as alternative agents to halon.

Dry chemical and wet chemical systems are used mainly for restaurant hoods, ducts, and cooking appliances found in kitchens and cafeterias. Halon, halon replacements, and, in certain situations, carbon dioxide are used in electrical switchgear rooms and in computer and data processing installations. In a building equipped with sprinklers, the cost of separate fixed systems and the estimated cost of likely damage in the event of activation of that type of system should be compared with the cost of sprinkler protection and the likely cost of sprinkler system water-related damage.

Smoke Control

Smoke can be hazardous to people and property, including a building itself and its contents. One obvious way to control smoke is to limit the use of flammable synthetic materials in modern furnishings and furniture. Such materials should, if possible, be kept to a minimum, and only those of a fire-resistive quality should be permitted in buildings. Of course, the proliferation of personal computer systems in the workplace has made the limitation of combustible materials all the more difficult. Smoke control measures are affected by the design and construction of a building. The *Fire Protection Handbook* notes:

> Smoke can behave very differently in tall buildings than in low buildings. In low buildings, the influences of the fire, such as heat, convective movement[s], and fire pressures, are generally the major factors that cause smoke movement. Smoke removal and venting practices reflect this behavior. In tall buildings, these same factors are complicated by the stack effect, which is the vertical natural air movement through the building caused by the differences in temperature and densities between the inside and outside air. This stack effect can become an important factor in smoke movement and in building design features used to combat that movement.
>
> The predominant factors that cause smoke movement in tall buildings are stack effect, the influence of external wind forces, and the forced air movement within the building.[19] (Milke, 2003, p. 12-116)
>
> Forced air movement is caused by a building's air-handling equipment and air-conditioning and ventilating systems, "but it should be noted that air movement can be influenced significantly by the mechanical systems of the building" (Milke, 2003, p. 12-116).
>
> Heating, ventilating, and air-conditioning (HVAC) and air-conditioning and ventilating (ACV) systems are found in most high-rise buildings. As one writer observes, "Air-conditioning and ventilating systems, except for self-contained units, invariably involve the use of ducts for air distribution. The ducts, in turn, present the possibility of spreading fire, fire gases, and smoke throughout the building or area served" (Webb, 2003, p. 12-237). Therefore, "the location of equipment and the fresh air intakes, the types of air filters and cleaners, the system of ducts and plenums,[20] and the use of fire and smoke dampers[21] are crucial in limiting and containing a fire" (Webb, 2003, p. 12-242).

---

[19] The movement of smoke in high-rise buildings can also be complicated by post-construction building modifications that result in improperly sealed openings in floors. Called poke-throughs when they are small and installed to permit telecommunications or other wiring or ducting to pass between floors, such openings permit smoke to migrate. If such openings occur on enough floors, the building's air handling system can be impaired.

[20] A plenum is "a compartment or chamber to which one or more air ducts are connected and that forms part of the air distribution system. It can be used to supply air to the occupied area or it can be used to return or exhaust air from the occupied area" (Webb, p. 12-241).

[21] Fire and smoke dampers are installed in air-conditioning or ventilating ducts to automatically restrict the spread of fire and smoke, respectively.

It is not possible to thoroughly address such a complex subject as smoke control within the scope of this chapter, particularly as the operation of systems designed to control smoke movement vary substantially from manufacturer to manufacturer and building to building and also differ according to the laws, codes, and standards in effect at the time the system was installed. However, generally speaking, there are two basic approaches to the issue. To restrict the spread of fire and smoke throughout a building or area "the HVAC system can be shut down and the fire area isolated or compartmented.[22] Another approach is to allow fans to continue to run, using the air duct system for emergency smoke control" (Webb, 2003, p. 12-241). Moreover, in many high-rise buildings, "when a fire alarm occurs, there is automatic pressurization of stairwells using [mechanical] fans that keep smoke out of the stairwells" (Craighead, 2003, p. 207).

Fire Department or Emergency Services Notification

When an incident involving the building fire life safety system occurs, the fire department or emergency services must be notified immediately. In addition to notification by assigned staff at the building, monitoring of the fire life safety system by an off-site central station or by the fire department itself may be required by local code or may be an additional safety measure taken by the building owner and manager.

Occupant Notification

When a fire or fire alarm occurs in a high-rise structure, it is critical that all affected occupants and responders be notified promptly. As the *Fire Protection Handbook* notes:

> Fire alarm systems protect life by automatically indicating the need for the occupants to evacuate or relocate to a safe area. They may also notify emergency forces or other responsible persons who may then assist the occupants or assist in controlling and extinguishing the fire. Audible and visual notification appliances alert the occupants, and, in some cases, emergency forces, and convey information to them. (Schifiliti, 2003, p. 9-35)

A fire alarm system that simply sounds an audible signal (in the form of bells, sirens, and whoopers) and flashes strobe lights is conveying a single bit of information: fire alarm. Systems that send voice announcements[23] or that use text or graphic annunciators typically convey multiple bits of information. They may signal a fire alarm and give a specific location and information on how and where to evacuate or relocate. When provided with detailed information about a fire emergency, people tend to evacuate more quickly and effectively. Audible and visible appliances "may also be used to indicate a trouble[24] condition in the fire alarm system, or they may be used as supervisory signals to indicate the condition or status of other fire protection systems, for example, automatic sprinklers" (Schifiliti, 2004).

---

[22] "Barriers such as walls, partitions, floors, and doors with sufficient fire resistance to remain effective throughout a fire exposure have a long history of providing protection against fire spread. These same barriers provide some level of smoke protection to spaces remote from the fire" (Webb, 2003, p. 12-120). Compartmentation also includes the automatic closing of fire doors, such as elevator or lobby doors.

[23] Announcements may be live or prerecorded, depending on the requirements of the local jurisdiction.

[24] A trouble signal is "a signal initiated by the fire alarm system or device indicative of a fault in a monitored circuit or component" (*National Fire Code,* 2001).

A public address (PA) system is a one-way system that allows voice communication from the building's fire annunciator and control panel (located in many high-rise structures in a room known as the fire command center or fire control room) to occupants of the building.

Emergency Planning

To adequately address the life safety of a high-rise structure, it is essential that a comprehensive emergency management plan be created. An emergency management plan describes the actions to be taken by an organization to protect employees, the public, and assets from threats created by natural and man-made hazards. In developing an emergency management plan, managers anticipate possible threats and make all the initial decisions ahead of time so that, in an emergency, they can focus their time and attention on the most important actions required.

For a high-rise structure, the plan should address such issues as the following:

- posted evacuation signage that clearly marks the means of escape from areas within the building

- description of the types of building emergency systems and equipment and how they operate

- the nature of the building emergency staff organization (including building management and the building fire safety director; engineering, security, janitorial, and parking staff; and floor response personnel such as floor wardens, or fire wardens, and other necessary persons, including those responsible for assisting disabled persons) that will handle emergency response until outside response agencies arrive

- contact details for persons and agencies needed in an emergency

- procedures for building emergency staff to handle each emergency expected for the building, including the methods of evacuation and relocation[25] and outside the structure

- how building occupants, floor wardens and building emergency staff are trained in life safety, including the frequency and nature of evacuation drills—commonly known as fire drills

Of course, the plan must be kept up to date in order for it to be effective.

---

[25] The need to evacuate or relocate occupants depends on the nature of the emergency. For most high-rise emergencies, it is not necessary to evacuate occupants to the outside of a building but rather to conduct a staged evacuation whereby they are relocated to an area of safety within the building; nor is it necessary, except under extreme circumstances, to conduct a total evacuation of all building occupants.

HIGH-RISE STRUCTURES

3.3. SECURITY CONSIDERATIONS

3.3.1. SPECIAL CONCERNS OF HIGH-RISE STRUCTURES

Previously in this chapter the special concerns of high-rise structures from a life safety perspective were addressed. Those concerns from a security perspective will now be looked at. As one author observes:

> From a security perspective, high-rise office buildings differ from low-rise buildings in two ways. First, the existence of multiple, occupied floors, one on top of another, means a higher concentration of occupants and therefore more personal and business property to be damaged or stolen compared with that in low-rise buildings. The potential for theft is increased by the fact that the concentration of property makes the site more attractive to a criminal; also, the greater the concentration of people, the better the chances of the thief's anonymity. Second, the more individuals assembled in one location at any one time, the higher the possibility is of one of these persons committing a crime against another. One of the difficulties in making the aforementioned statements is the lack of crime pattern analyses[26] is a high-rise or low-rise, largely depends on the design and construction of the building itself, the neighborhood in which it resides, and the security program that is in place. (Craighead, 2003, p. 27)

The number of occupants and visitors in a high-rise structure can be high.[27] Access to and egress from such buildings is often funneled through the core of the structure where vertical transportation, such as elevators and escalators, is provided. Authentication and control of access and egress requires special consideration in case of a terrorist threat or other criminal acts, such as the theft of personal or business property (like portable computers) or acts of workplace violence. The development and application of meaningful and effective security measures for the high-rise building environment helps to minimize these and other risks.

The Problem of Exposed Assets

Significant assets may be at risk in multi-tenant properties that house banks, retail stores, brokerage firms, professional offices, and headquarters of large organizations. As the ASIS International Commission on Guidelines has observed:

> *Assets* include people, all types of property, core business, networks, and information. *People* include employees, tenants, guests, vendors, visitors, and others directly or indirectly connected or involved with an enterprise. *Property* includes tangible assets such as cash and other valuables [e.g., furniture, fixtures, and art objects] and intangible assets such as intellectual property and causes of action. *Core business* includes the primary business or endeavor of an enterprise,

---

[26] "*Crime pattern analysis* is a process that encompasses a number of techniques, all of which can assist crime risk management. It is therefore best regarded as a generic term, covering a number of approaches and techniques for analyzing the incidence and distribution of crime" (Kitteringham, 2001, p. 12).

[27] For example, according to a Port Authority of New York and New Jersey briefing (May 13, 2004), "On any given workday, up to 50,000 office workers occupied the [New York World Trade Center] towers, and 40,000 people passed through the complex" (*9/11 Commission Report*, 2004, p. 278).

including its reputation and goodwill. *Networks* include all systems, infrastructures, and equipment associated with data, telecommunications, and computer processing assets. *Information* includes various types of proprietary data. *(General Security Risk Assessment Guideline,* 2004, p. 6)

Many of these assets are vulnerable not only to natural hazards but also to theft and malicious damage. In buildings open to the public, the often large number of people moving in halls, corridors, elevators, stairs, and public spaces makes the assets' vulnerability even greater.

For the purposes of high-rise protection, it is not relevant that the risk of loss be shared by various organizations. The need for protection flows from the nature of the assets, not from the identities of the owners. If a tenant with a high exposure chooses not to protect its assets, there may be a threat to another tenant because the unprotected assets could attract attackers to the building. The final protection scheme should involve a cooperative effort between the building owner and manager and the tenants.

3.3.2. LIFE SAFETY AND SECURITY DILEMMA

The factors that complicate life safety in a high-rise structure can also create security problems. For example, code provisions that require unimpeded exit capabilities at all times when a building is occupied mean that egress may be possible with little or no surveillance. During a genuine emergency, security forces may be unprepared for the joint demands of emergency response and heightened security.

Furthermore, some life safety requirements make it possible for two persons to stage an event that could render a target location within the building vulnerable. One person could simulate a fire emergency on one level by activating a manual fire alarm, and an accomplice, many levels away at the target location, could take advantage of doors unlocked due to the emergency. To maintain security under such conditions, countermeasures can be used. These include intrusion alarms and closed-circuit television (CCTV) featuring alarm-triggered camera display and video recording.

The following sections address available security resources and their appropriate applications in protecting high-rise structures. A typical high-rise building is described below for purposes of illustration.

3.3.3. TYPICAL HIGH-RISE

The main ground level lobby of a typical high-rise structure is illustrated in Figure 1. Profile data for the fictitious building are as follows:

HIGH-RISE STRUCTURES

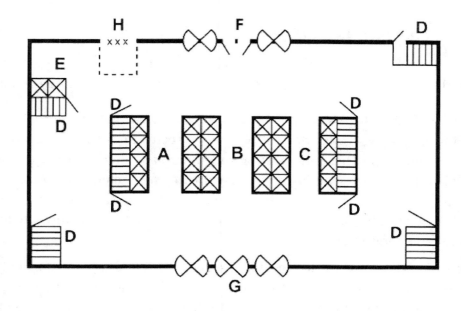

FIGURE 1
High-Rise Office Building General Lobby

Legend:
A = High-rise elevator lobby (eight cars serve)
B = Mid-rise elevator lobby (eight cars serve)
C = Low-rise elevator lobby (eight cars serve)
D = Fire stair enclosures and doors
E = Service elevators (two cars serve all floors B2 through 40)
F = Rear doors with one set double swinging and two revolving
G = Main doors with three revolving and two single swinging
H = Overhead door and ramp down for trucks

- The building is 40 stories high.

- It has about 6,000 occupants divided into roughly 150 occupants per floor.

- The passenger elevators in each of the three banks serve approximately one-third of the building, while the service or freight elevator cars serve all floors, from the basement levels through 40.

The two basement levels, B-1 and B-2, are mechanical and storage spaces served by the two service elevator cars and by the low-rise bank (C bank) elevators. Level B-1 features a vehicle entrance via the ramp shown at H in the figure. The ramp entrance is secured at the street level by an overhead, metal roll-down gate and at the bottom of the ramp by an overhead, open-grille roll-down gate. Inside the area protected by the latter is a loading dock that can accommodate four trucks and is served by the service elevators.

3.3.4. BUILDING OPERATING MODES

Before discussing security measures, it is helpful to understand some common security concepts for high-rise structures. The first is the concept of "open" or "closed" buildings:

1. *"Open" buildings.* Access is typically unrestricted at the building entry level. Tenants, building employees, and visitors proceed directly to their destination floor, via building elevators, where the occupier of that space determines the level of entry control and security. Open buildings may provide a concierge or security desk at the street level to provide directions and deter access by "undesirables" (if they are easily identifiable as such) and to implement security procedures for messengers, delivery persons, and contractors.

2. *"Closed" buildings.* Access to elevator banks is controlled by some form of credential checking, visitors are authenticated prior to proceeding to elevators, and sometimes a separate messenger center, with controlled delivery of property, is provided.

3. *"Hybrid" buildings.* Where a major tenant may occupy a large part of the building, one or more elevator banks may be designated as closed, while other elevator banks that provide access for multiple tenants are operated in an open mode.

The second concept is that buildings operate in different modes depending on traffic, occupancy level, hour of the day, and day of the week:

1. *Business hours.* The regular mode of building security operates during the regular office hours for the majority of the tenants.

2. *Intermediate hours.* These are usually an hour or two before and after regular business hours and possibly Saturdays, when occupancy is light and enhanced security is required. The building main entry doors may remain unlocked, but tenants and visitors may be more closely scrutinized.

3. *Off-hours or after-hours.* During nighttime, weekends, and holiday hours, when fewer tenants are present, enhanced security is required, particularly when maintenance and cleaning crews are working and special activities (such as tenant deliveries and tenant space modifications) are in progress. The building perimeter main entry doors are usually locked, and entry requires a credential, such as an access control card, or the summoning of a building employee or security officer by a signaling device (such as an intercom or buzzer).

3.3.5. BUILDING ELEMENTS

Every high-rise structure has one or more highly security-sensitive locations; some buildings may be sensitive throughout the entire structure.

The classical approach to perimeter security views a property in terms of rings. The property boundary is the first ring, the building is the second, and the specific interior spaces are the third (Healy, 1983, Chapter 3). This scheme needs only a slight variation to fit the high-rise structure: The building line is usually the first ring since, in an urban environment, the building line is often adjacent to a public sidewalk; access to vertical transportation (stairs, escalators, and elevators) is the second ring; and individual floors or floor sections are the third.

For access control purposes, it is helpful to divide the high-rise structure into three classes or types of interior spaces:

1. *Public access or common areas.* These include street-level entrance lobbies, main elevator lobbies, access routes to retail spaces and restaurants in the structure, promenades, mezzanines, and, increasingly in new buildings, atria.

2. *Rented or assigned occupancies (i.e., tenant spaces).* These areas are leased or owner-occupied spaces on various floors. Depending on the occupant, such spaces may be open to public access during business hours or may be restricted to identified and authorized persons.

3. *Maintenance spaces.* These include mechanical rooms and floors, communications and utilities access points, elevator machine rooms, janitorial closets, and spaces with strict limited access.

Access control measures are different for each class of space, depending on the building security program.

3.3.6. ACCESS CONTROL OF PUBLIC AREAS

In order to control access of persons entering the main lobby of the building, such persons need to be categorized. For example, regular tenants may be issued a credential—such as an access control card—that permits rapid verification of their right to enter. However, if visitors need to be authenticated, communication with the visitor's host may be required. The table below (Figure 2) illustrates some of the categories of persons entering the building and the areas to which they may require access. For areas noted with "?," access depends on building operations and the security level required.

Traffic Category	Retail Space	Loading Dock	Pass'ger Elevator	Service Elevator	Tenant Space	Fire Exits	Mech., Elec. Mach. Rms	Roof
Retail Tenant	•	•		?		•		
Retail Customer	•					•		
Office Tenant	•		•		•	•		
Office Visitor	•		•		•	•		
Contractor	•	•	•	•	•	•	•	•
Building Maintenance	•	•	•	•	•	•	•	•
Delivery Driver	•	•	?	?	?	•		
Messenger	•	•	?	?	?	•		
Janitorial Crew	•	•	?	•	•	•		

FIGURE 2

Traffic Type & Access

A possible control scheme for a building is that access to all floors above the main lobby is controlled by locking the building stairwells or fire stairs against entry from the lobby while allowing free egress to the lobby and placing control points at each bank of passenger elevators. To be completely effective, such a scheme would require that all persons using the elevators to any level present some form of personal identification or permit inspection of inbound property, or both. During periods of heightened security this may be acceptable. However, under normal circumstances, building tenants may resist such tight security. This is especially true of tenants with walk-in customers. If the building contains a bank, it will generally be located on the ground level and perhaps use part of a basement as well (for its vault and safe deposit areas). Bank lobby access will have to be fairly unrestricted.

If retail operations are restricted to the ground floor and separate street access can be provided, upper-floor security need not be compromised. (Such an arrangement is particularly important for a closed building that houses, for example, a cafeteria that is open to the public. It is important that visitors be able to enter the cafeteria freely but not have access to the rest of the building.) If, however, an owner occupies some floors and rents others, it is important to plan the space allocation or stacking arrangement of tenants to facilitate access control at a minimum cost and inconvenience. If, for example, the owner plans to occupy half the building and rent the other half, or make any other allocation between proprietary and rental floors, the building can be operated as a hybrid, and the following security planning points should be considered:

HIGH-RISE STRUCTURES

1. *Identify particularly sensitive owner occupancies.* These include executive offices, data processing facilities, cash or securities handling, activities involving large amounts of sensitive proprietary data, and any other space that the owner has designated as critical.

2. *Isolate sensitive owner spaces.* To the degree possible, sensitive spaces should be grouped together on floors that constitute a single sector (low-rise, mid-rise, or high-rise) of the structure.

3. *Impose access controls on the building sector containing the sensitive occupancies.* In the typical high-rise, assume that the owner has chosen to occupy the floors served by the high-rise (bank "A") plus half of the mid-rise (bank "B") elevators, and to rent the others. The sensitive occupancies could be stacked in the floors served by the high-rise bank. If, in the occupancy of the mid-rise sector, the division of floors is about equal, the mid-rise lobby could be divided in half and the mid-rise elevators programmed so that half the cars serve floors 13 through 19, the other half floors 20 through 26. The half serving sensitive occupancies (say, floors 20 through 26) would then be controlled in the same way and probably using the same security personnel controlling the high-rise bank, using card readers and line or other traffic-flow devices. Should the occupancy ratio change in the mid-rise, the controls could easily be abandoned without affecting the high-rise element.

4. *Ensure that controls cannot be bypassed.* If the elevators have been designed so that certain floors are common or "crossover" for two or more elevator banks, it is important to ensure that transfer at such a floor will not permit uncontrolled access to a sensitive area. For example, if floor 27 is served by both the high-rise and the mid-rise cars to permit crossover on that floor, it would be necessary either to install an access checkpoint using security staff on floor 27 or to control access to sensitive floors using card readers in the elevator cars that serve both banks. If no crossover floor is provided, persons who wish to go from the high-rise to the mid-rise or vice versa must descend to the main lobby to transfer to the other elevator bank.

Elevator Control

Throughout the life of a high-rise structure, changes in occupancy may require changes in access control requirements. It is a major advantage if the passenger elevators are originally designed (or retrofitted when that would be cost-effective) to permit access control of all floors. With such control, it is possible to program each car individually as to how it will respond to calls from within the car and from elevator lobbies. Also, the installation of card readers in all elevator cars, even if the readers are not activated at all times, is a sound security measure. An additional security measure to screen persons before they enter an elevator is the use of turnstiles. If, in the future, due to a change in security needs, the building must move from being an open building to being a closed building, the transition can be achieved by programming the elevators to be on card access control 24 hours per day, seven days per week, and issuing access cards to those persons authorized to access designated floors for time periods determined by tenant management.

Service or Freight Elevators

Service or freight elevators pose special problems in all high-rise structures because they often serve all levels of a building. If they are self-service, the entire building security program may be compromised unless the service cars are

- programmed not to access sensitive floors without special arrangements,
- locked at hoistway doors on sensitive floors, or
- locked at service vestibule or lobby doors, where applicable.

Moreover, service cars offer an opportunity for unauthorized movement of property from or between accessible floors. A preferable approach is to assign an operator to the service cars or make them available only on request to the security operation. It may also be useful to install electronic card readers in the elevator cars and CCTV cameras in the cars or elevator lobbies. In any event, use of service cars for general passenger movement should be avoided.

Building Stairwells

Building emergency exit stairwells (building stairwells) or fire stairs are part of the public access or common areas and must be accessible for occupants to escape[28] (particularly when building elevators are not available or are unsafe to use, such as during a fire emergency).

Two factors control how building stairwells are secured. One is local fire and building code requirements. During times of occupancy, building stairwells must never be locked in the path of egress. The code may, however, allow restrictions on reentry from the stairwell. For example, some jurisdictions permit doors leading from the stairwell to the floor to be locked as long as on every fourth floor the doors are unlocked to allow an occupant to be able to leave the stairwell enclosure. Other jurisdictions permit locking of the stairwell doors as long as they automatically unlock (or fail safe) when the fire life safety system is activated. The other factor is whether inter-floor movement of building occupants via the building stairwells is permitted. Such movement may be argued for (particularly with multi-floor tenants) because it can save time and reduce demands on elevators and energy consumption, and it is convenient for building occupants to travel between floors. However, such an arrangement may detract from the security of the stairwells themselves. A better solution, costly as it may be, is to provide an internal staircase serving the needs of the multi-floor tenant. If use of the stairwells is permitted by the local jurisdiction, the

---

[28] Codes mandate that at least two emergency exit stairwells be provided per floor. This permits evacuating occupants to use a second stairwell if their nearest one is unusable due to emergency conditions.

stairwell doors need to be equipped with approved access control devices, such as card readers, as well as intercoms to permit two-way communication by occupants with building staff. The selection of locking devices for high-rise building stairwell doors is critical if code requirements are to be met. Any such arrangements must be closely coordinated with the local authority having jurisdiction.[29]

The following quote emphasizes the challenge of stairwell exit doors:

> The fact that security and fire life safety are different disciplines, and that their priorities are sometimes in conflict with each other, is nowhere better demonstrated than at the stairwell exit door.... [T]he need to maintain immediate, [unhindered] exit from the stairwell at the ground level provides an opportunity for a person who has perpetrated a crime within the building to make a rapid exit. (Craighead, 2003, pp. 130-131)

To address this issue, some authorities allow installation of a delayed egress locking device that permits, during non-emergency times, a 15- to 30-second delay in opening the emergency exit fire door from the inside of the stairwell. Such a lock can be helpful when used in conjunction with the following:

- an alarm that notifies security staff that a person is trying to exit the location,
- an intercom to communicate with the person, and
- a CCTV camera to view and record the event.

The time delay may be sufficient for security staff to intercept the person. These locks are required to unlock automatically if there is a loss of power to the lock or the building fire life safety system is activated.

Intrusion alarms (explored later in this chapter) are important on stairwell doors. The alarms indicate entry into the stairway and permit a response. Where stairwell doors are generally locked against floor reentry, an appropriate response would be to dispatch a security officer to the ground level door in the stairwell involved and to send a second officer to enter the stairwell at a point above the entry point indicated by the intrusion alarm. That response would permit interception of the intruder while the person is still in the stairwell enclosure. For stairwell doors that are generally unlocked, intrusion alarms on sensitive floors indicate use of the fire stairwell door on those floors. Use of stairwells is usually prohibited, except in genuine building emergencies. Specific response can then be made to the floor when an alarm has been set off. To function properly, building stairwell intrusion alarms (in fact, intrusion alarms anywhere) must be zoned to indicate precisely which door or area has been penetrated.

---

[29] The "authority having jurisdiction" is defined as "the organization, office or individual responsible for approving equipment, materials, an installation, or a procedure" (*National Fire Code*, 2001).

3.3.7. ACCESS CONTROL OF INTERIOR FLOORS AND SPACES

The preceding discussion dealt with public access areas; however, many floors and interior spaces within floors also require access control. Such control may be the only control measure, or it may be practiced in addition to general control at the building lobby level.

Programmable elevators permit some floor control. A variant of that approach, even when total floor control is not possible using the elevator system, is to install card readers in the elevator cars to control access to certain floors. Visitors and others without access cards will then require special handling by building security staff. On a floor with few visitors, an escort could provide the required security. On a floor with heavy visitor traffic, a better solution might be to allow free access via elevators serving that floor, and then to provide a floor reception point to screen admittance. In addition, controlling elevator use with access cards requires authorized users to help deter piggybacking or tailgating, whereby an unauthorized person gains entry by accompanying an authorized card holder to a secured location. Such activity can also be controlled using optical turnstiles near elevators to screen persons before they proceed to an elevator car.

Depending on design, floor reception can often be achieved at the elevator lobby itself. In the building shown in Figure 1, one end of the elevator lobby on an upper floor would need to be locked against entry to the floor (perhaps with provision for remote latch release or card entry), and the reception control could be established at the opposite end of the elevator lobby.

In a major integrated security system, staff at the central console could readily exercise remote access control for one or more floors or areas within floors. Access to smaller spaces within individual floors can be controlled with conventional keys, card readers, remote latch release, and communications/surveillance capability, or by a receptionist.

Special consideration must be given to the use of locking devices on doors leading to elevator lobbies on floors above ground level. For example, in many buildings, during a fire emergency—in particular, the activation of an elevator lobby smoke detector, an elevator hoistway smoke detector, or an elevator machine room smoke detector—the elevator cars serving the affected elevator bank automatically return to a designated level or to an alternate floor (if the activation of the elevator lobby smoke detector occurred on the designated level).[30]

---

[30] "*Designated level* is defined as the main floor or other floor that best serves the needs of emergency personnel for firefighting or rescue purposes" (Donoghue, 2003, p. 12-203). "The automatic or manual return, or recall, of elevators to a designated level or alternate level" (Donoghue, 2003, p. 12-202) is known as Phase I operation. Once the elevators have been recalled, they are then available for firefighters' service using a special key. "The provision that allows firefighting personnel to operate the elevator from within the car, or emergency in-car operation, is commonly referred to as Phase II" (Donoghue, 2003, p. 12-202). Since this use of elevators is for fire department personnel only, not building engineering or security staff, the use of such a key should be strictly controlled and restricted to the purpose for which it was designed.

Therefore, the selection of any locking device for elevator lobby doors is crucial because any persons who may be in the lobby when the elevators recall must be able to exit the elevator lobby and proceed to a building emergency exit stairwell. One should check with the local authority having jurisdiction regarding the types of locking devices that are permitted on these doors.

3.3.8. ACCESS CONTROL OF BUILDING MAINTENANCE SPACES

Access to critical areas such as mechanical rooms and floors, air-conditioning rooms, fire pump and fire control rooms, telecommunications and utilities facilities (including electrical transformer vaults), elevator pits, and elevator machine rooms should be controlled using selective admittance devices and monitored using intrusion alarms. Each controlled door, if opened other than with the use of an authorized key, automated access device, or remote latch release, would send a monitoring location a distinctive intrusion alarm designed to initiate a response to the intrusion.

Even in large buildings with multiple maintenance spaces, it is feasible to maintain access control. The limited population of persons requiring access should be readily identified. In a fully automated system, such persons will use access cards and readers or be remotely admitted using remote latch control, CCTV surveillance, and two-way communication. Even in a smaller structure for which a major integrated security system investment is not warranted, a minimum, low-cost control would involve the following:

- conventional lock on area door

- controlled key issuance

- intrusion alarm, distinctively zoned

- telephone or radio contact by an authorized key holder immediately before or after entry (the need for communication could be eliminated by equipping the intrusion alarm with a shunt switch operated with an authorized key)

The disadvantage of this approach is that keys can be lost and compromised.

Utility personnel require access to some of these spaces, particularly telephone installers to telephone frame rooms and terminal panels. The problem can still be handled without surrendering security control. An installer, regularly assigned to the structure, can be issued a key or automated entry device. For a temporary worker, after identity has been verified, a one-day-only key or device can be issued; however, its return must be documented before the worker leaves the premises.

3.3.9. ACCESS CONTROL OF AIR INTAKES AND TELECOMMUNICATION SERVICES

Access to fresh air intakes and telecommunication services should be protected.

Fresh Air Intakes

Fresh air intakes for HVAC systems are often located at or near ground level, where they are usually covered with louvered intake screens. These should be protected against introduction of airborne chemical or biological agents. Figure 3 suggests an approach using a grille or baffle at 90 degrees to the actual air intake with an intrusion alarm on the grille. Admittedly, the intake suction could still draw vapors from a source placed on the grille. An even better solution is to locate such air intake ports sufficiently far above the ground to make access impossible.[31]

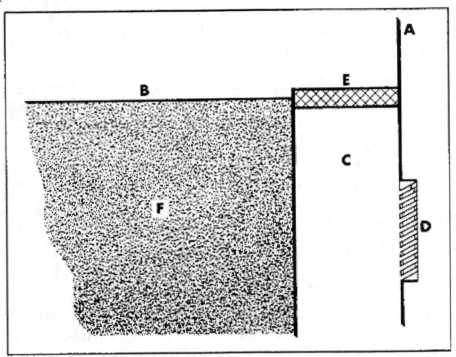

FIGURE 3
Fresh Air Intake Protection Scheme

Legend:

A Building exterior wall
B Grade level
C Open space and intake area
D Louvered intake shutter
E Grille (to be protected with intrusion alarm)
F Unexcavated earth

---

[31] Ways to protect vulnerable outdoor air intakes are documented in *Guidance for Protecting Building Environments from Airborne Chemical, Biological, or Radiological Attacks*, Publication No. 2002-139. by NIOSH (National Institute for Occupational Safety and Health), Department of Health and Human Services, Centers for Disease Control and Prevention, May 2002, www.cdc.gov/niosh.

HIGH-RISE STRUCTURES

As well as protecting air intakes against unauthorized access, security measures such as intrusion detection devices, CCTV, and security personnel should be used to uncover any security breaches. Also, provision for automatic or manual shutoff of the HVAC system should be made. For a quick response, any manual shutoff device should be readily accessible to staff (such as building engineering or security) authorized to perform the shutoff.

Telecommunication Services

Telecommunication services typically leave a building in one or more trunk cables and connect with the distribution network in the street conduit and tunnels. Telephone, computer, and alarm systems all use the telecommunications path and are vulnerable to attack along the telephone cable system. For a high-security building, two physically separated paths for telecommunications cable to separate telephone centers are recommended.

The most dangerous or critical points for attack are located at points after all the building services are connected with one or two main trunk cables still in the structure. It is not uncommon in older structures to find such trunk cables exposed in basement or utility area spaces inside the building. In newer structures, greater care should be taken to conceal the cables.

It is important to make a careful check of the main cable routes from the communications risers in the building back out through the service entrance. Any access points along the route should be protected. If the cable is in an enclosed space, then intrusion detection devices are required for access points such as hatch covers and doors. If the cables are exposed, consideration should be given to hardening the cable path by resistive construction.

3.3.10. SECURITY FEATURES

A number of security features can be applied to high-rise structures. These include lighting, selective site hardening, building locks and locking devices, alarm sensors, duress alarms, CCTV and integrated security systems, security personnel, and the policies and procedures used to safeguard the assets of a building. More information on these subjects, beyond what follows, is found in other chapters of the *Protection of Assets Manual*.

Lighting

Lighting is an important feature in providing protection for high-rise structures. It is a deterrent to criminal activity, including theft and physical assaults. It can be used for perimeter approaches (such as roadways and pedestrian walkways), perimeter barriers, site landscaping, building facades, open parking areas and parking structures, and all areas within a building, including loading docks. Lighting also aids in the effective use of CCTV systems.

Another chapter of this manual provides a basic understanding of lighting, the features and benefits of different types of lamps, and the effective application of lighting to increase safety and security.

Selective Site Hardening

An individual or company that is already a high-risk target for criminal activity, including terrorism, may be further exposed in a high-rise structure. In addition to the access control efforts already addressed, there may be a need for specific hardening of a structure at or within particular target floors. Ideally, such security measures should be considered during initial construction, since options will be fewer in a retrofit.

Construction Features. Resistivity to impact and firearms is a principal reason for special construction. (However, terrorist bombings have added the need to evaluate blast-resistant design for new, high-profile construction.) Materials capable of providing resistivity include standard masonry, blast-resistive reinforced concrete, sheet steel, polycarbonates, acrylic material (Plexiglas), compressed fiberglass, Kevlar and ballistic fabrics, bullet-resistive glass, and security window film (to hold broken glass firm, preventing its shards from becoming lethal projectiles). Another chapter of the *Protection of Assets Manual* discusses glazing materials, their resistance to projectiles and bombs, and their reflective properties.

A combination of these materials can be aesthetically integrated into design plans to provide protection for the designated space. For example, on a floor occupied by key personnel in the building illustrated in Figure 1, both ends of the elevator corridor could be blocked off with resistive material and provided with controlled doors of the same or equivalent material. The partition could be made of masonry and fiberglass, or of steel between conventional drywall panels, and the door could be constructed of resistive glass or polycarbonate. It is less expensive and less confining to total floor architecture if control barriers and hardened materials are used as close to the normal points of floor entry as possible. Hardening and controlling the elevator lobby and all stairwell doors is less expensive than hardening individual offices or spaces within the floor.

Because many high-rise buildings feature dropped ceilings with air-conditioning plenums or ducts that allow ample space for intrusion, it is important when hardening selected floors to pay attention to the ceiling (or below-the-floor spaces). Protection can be achieved by continuing the resistive partitioning to the floor slab above or using heavy gauge wire mesh to close the space but still permit air movement. Intrusion devices are recommended in conjunction with such ceiling protection.

Key Reception Area. If entry control for a target floor is on the floor itself, the reception point must be resistive to attack. However, executives often resist protective barriers around executive floor reception points, saying such barriers set the wrong mood or give a poor impression. Several solutions are available. One is to establish reception on a floor other than the protected floor—perhaps a floor above or a floor below. At that reception point, the visitors can be identified and their presence communicated to a control person on the sensitive floor. Arrival at that floor, which is usually locked against entry, is expected, and as soon as the announced parties arrive at the elevator lobby entry point, the control person can unlatch the door and admit them.

Another approach is to create two control points on the protected floor. The outer or reception point would be staffed and set off from the interior floor space by protective barriers in or through which an observation capability would permit a person inside or behind the protected barriers to unlatch the barriers to admit an authorized visitor to the inner area. The guests or visitors would not be aware that the executive floor inner doors were locked against them. The unlatching could be in advance of their movement from the reception desk to the door. Although the receptionist would still be potentially exposed, absence of an unlatching mechanism at that point would prevent access. One would either have to be admitted by the person inside the protected area or carry a key or other admittance device to open the door from the reception side. The receptionist should not possess such a key.

Physical Barriers. Various physical barriers can be used to protect structures against such threats as vehicle bombs. These include perimeter fences, walls, sidewalks, pathways, landscaping, fountains, pools, sculpture, benches, planters, barricades, and bollards. Another chapter of this manual examines structural barriers and the time it may take to breach them. The difficulty is to achieve sufficient spatial separation between the barrier and the structure itself. Since many high-rise structures are located in major cities and towns, where space is at a premium and the structures already exist, this goal is often unachievable. However, properly designed perimeter barriers can keep a vehicle containing explosives from being driven into the structure itself, prior to its detonation.

Parking controllers and barriers can be used to control the entry of vehicles into parking garages and loading docks, particularly those located under buildings.[32] Gate arms and, in higher-security applications, crash-rated hydraulic impact barriers can be manually or automatically activated to restrict entry to authorized vehicles only.

---

[32] Unfortunately, due to the absence of sufficient parking spaces in major metropolitan areas, high-rise structures, particularly newly constructed ones, often need to provide under-building parking facilities.

Building Locks and Locking Devices[33]

High-rise structures use many types of locks and locking devices. The type of locking device used at any particular location depends on the type of door or barrier on which it is installed and the prevailing building or life safety code requirements. The following are several types of applications and appropriate locks:

- **Building main entrance/exit doors and perimeter legal exit doors: electromagnetic locks.** Using a timer, these locks can be designed to automatically lock and unlock to accommodate the operating requirements of a building. As one source notes, "When the doors are secured, access from the outside of the building can be obtained by hooking up a card reader (normal egress is permitted using panic hardware[34] or automatic unlocking devices such as motion detectors). . . . Life safety codes mandate that the power source to all locks restricting occupants' means of egress must be supervised by the building fire life safety system. In the event of an emergency, such as a power failure or activation of a fire alarm, electrical current to the electromagnet ceases and the doors unlock. Occupants can freely exit the building and responding emergency agencies, such as the fire department, can enter" (Craighead, 2003, pp. 127-128).

- **Inside of stairwell doors: a special hybrid electric locking device known as a high-tower-function mortise[35] lock or stair tower lock.** Such locks were developed so a stairwell door can be locked on the stair side to provide security. A source notes, "Generally, high-tower-function mortise locks are energized and locked at all times. Access control is accomplished by either a mechanical key, digital keypad, or a card reader. . . . The power source for these locks is controlled by the building life safety system so that in an emergency, doors immediately unlock yet remain closed and latched, protecting the stairwell from smoke and fire" (Geringer, 1991, p. 2). Also, as was mentioned earlier, some authorities allow installation of a delayed egress locking device that permits, during non-emergency times, a 15- to 30-second egress delay in opening the emergency exit fire door from inside the stairwell. Such locks are required to unlock automatically if there is a loss of power to the lock or the building fire life safety system is activated.

- **Exterior of stairwell doors where they exit the building: key-operated pin tumbler locks and electromagnetic locks.** Stairwell exit doors that normally exit at the ground level "may be locked on the exterior side as long as long as, at all times, the inside of the door is operable, providing uninhibited egress. When an occupant applies pressure to

---

[33] Another chapter in this manual provides a general description of the different types of locking mechanisms and their relative strengths and weaknesses, as well as information about the specific locks (and locksets) to use for particular applications.

[34] "Panic hardware devices [such as cross bars or push pads] are designed to facilitate the release of the latching device on the door when a pressure not to exceed 15 lb. (6.8 kg) is applied in the direction of exit travel" (Lathrop, 2003, p. 4-72).

[35] A mortise lock is one where the lock is set into a mortise or recess in the door itself.

the fire exit hardware,[36] the door will immediately unlock [the exception being delayed egress locks]" (Craighead, 2003, p. 131).

- **Elevator lobby or vestibule doors: electromagnetic hold-open devices.** On activation of the building life safety system, electrical power to each hold-open device ceases, causing it to release the lobby door, which is then free to swing shut and protect the lobby against the intrusion of fire and smoke.

- **Doors leading to tenant spaces: key-operated pin tumbler locks and card access control devices.**

- **Doors inside tenant spaces that do not lead to a legal fire exit: key-operated pin tumbler locks, card-operated locks, token-operated locks, biometric identification system-operated locks, electrified locking mechanisms, and combination locks.**

- **Doors leading to building maintenance areas: key-operated pin tumbler locks, card-operated locks, token-operated locks, biometric identification system-operated locks, electrified locking mechanisms, and combination locks.**

Pin tumbler locks are found throughout high-rise structures. As one source notes:

> The pin tumbler mechanism is the most common type of key-operated mechanism used in architectural or builders' (door) hardware in the United States. The security afforded by this tumbler mechanism ranges from fair in certain inexpensive cylinders with wide tolerances and a minimum of tumblers to excellent in several makes of high-security cylinders. (Edgar, 2004, p. 168)

Pin tumbler locks can be master keyed. The principle of master keying is that a single lock may be operated by more than one key by designing various keys to engage or work on different tumblers or different aspects of the same tumblers. Master keying provides a hierarchy of access to groups of locks, from access to only one lock to access to all locks in the population.

Another chapter of this manual addresses general locking concepts and the recommended methods of maintaining acceptable levels of lock protection. With those principles in mind, the following suggestions are made for the overall lock program of a high-rise structure:

1. *Use interchangeable or removable core locks.* These permit frequent and rapid changes without floor or shop labor. Adequate planning assures that sufficient replacement cores in the various control and master groupings are on hand for immediate use. If a locksmith is on staff, much of the core preparation can be done in-house.

2. *Severely limit the use of a single control or a grand master key.* The obvious danger is total compromise of the locking system through physical loss or loss of accountability

---

[36] "Panic hardware that has been tested and listed for use on fire-protection-rated doors is termed 'fire exit hardware'" (Lathrop, 2003, p. 4-72).

of a control or grand master key (or code) or a single core containing the control and grand master combinations.

3. *Use multiple controls and multiple masters.* The building can readily be divided into functionally different units. For example, the low-rise, mid-rise, and high-rise floors can be regarded as separate locking zones. Each zone can then be set up to its own control and its own master scheme.

4. *Establish multiple key blanks.* In setting up different control and master schemes for major building elements, the key blank style (the profile of the key and keyway) should be different for each of them so that a blank for use in one element will not enter the keyway in the other elements. Such an arrangement helps prevent unauthorized key use. It also makes the same lock combination codes used in one sector available for use in another with low risk of compromise because the key blanks would not be interchangeable.

5. *Set up single zones.* Any distinction by floor or occupancy, or both, will serve the purpose. For example, if the mid-rise is a tenant section, then each of the floors could be provided a separate floor sub-master and all the floors could be included in a zone master. At the same time, spaces within the tenant zone not intended for tenant use, such as maintenance spaces, could remain on separate master systems that are building-wide but confined to a special function, such as electrical, mechanical, or janitorial.

6. *Maintain strict master key control.* Sub-master keys at the floor level should be strictly accounted for and issued only to senior personnel of single tenants. If several tenants occupy a single floor, none should have a floor master. Keys issued on this basis should be changed after changes in occupancy.

7. *Keep control of zone master keys.* These keys should be issued on a daily basis only, and only to persons with demonstrated, imperative need. Using multiple zone masters will reduce vulnerability to compromise but not eliminate it. Issuing zone master keys on a day-basis only (i.e., to be returned to security custody before the holder leaves the building) further limits exposure. Choosing very carefully those to be issued with zone master keys (or even allowed temporary custody of them) also narrows the exposure. The carrying of zone master keys as symbols of prestige should be actively discouraged.

8. *Limit access to control keys*[37] and combinations. Obviously, no one needs either unless a lock core is to be replaced or a new core pinned. The locksmith will have continuing access to both of these items, so the reliability of that person is of critical importance.

---

[37] Removable core cylinders use "a special key called the control key to remove the entire pin tumbler mechanism (called the core) from the shell [the shell is the cylinder housing]. This makes it possible to quickly replace one core with another having a different combination and requiring a different key to operate. Because of this feature, removable core cylinders are becoming increasingly popular for institutional use, and use in large commercial enterprises where locks are changed often" (Edgar, 2004, pp. 169-170).

Control keys should not be carried at any time except specifically when needed to make core changes.

9. *Pay particular attention to exterior doors.* These should not be included in any other building master or control system. As they are mostly on ground or below-ground levels, the doors[38] involved will typically lead to specialized maintenance spaces or to public access spaces controlled by the building owner or manager. Exterior doors which are locked against access to the building also require intrusion alarm protection.

10. *Maintain lock quality and complexity of combination schemes.* Although maximum security locks are probably not generally required for high-rise structures with a balanced system of security controls, the locks chosen should be of very good quality. Interchangeable core locks which use a single tumbler (pin chamber) to set up control combinations should be avoided because of the ease of creating a control key given any operating key and a little ingenuity.

Crucial to any locking system is an effective key control system:

> Before an effective key control system can be established, every key to every lock that is being used in the protection of the facility and property must be accounted for. Chances are good that it will not even be possible to account for the most critical keys or to be certain that they have not been copied or compromised. If this is the case, there is but one alternative—to rekey the entire facility. (Finneran, 2004, p. 193).

Depending on the size of the structure, that can be a very costly undertaking.

Alarm Sensors

Intrusion detection systems can be deployed in a high-rise structure for various perimeter, stairwell, maintenance, and tenant areas. For example, the following intrusion detection devices can be found in high-rise structures:

- *Passive infrared (PIR) sensors* are used to protect perimeters, areas, and objects; heating, ventilating, and air-conditioning air intakes; interior tenant areas; and building main exit doors (to facilitate automatic opening).
- *Magnetic contacts* are used on perimeter exit doors, building stairwell doors, doors leading to maintenance spaces, and doors inside tenant areas.
- *Break-wire sensors* are used on building heating, ventilating, and air-conditioning air intakes or in interstitial ceiling spaces.
- *CCTV motion detectors* are used in building stairwells and various locations where CCTV is used in the building.

---

[38] Typically, for security purposes, each emergency exit stairwell door will not have a locking mechanism on the exterior of the door.

Sensor technologies for the detection and annunciation of intrusion are examined in greater detail in another chapter of this manual.

Duress Alarms

A duress alarm (sometimes referred to as a panic alarm) is "a device that enables a person placed under duress to call for help without arousing suspicion" *(Threat Advisory System Response Guideline,* 2004, p. 8). Within a high-rise structure such alarms may be needed in banks and other financial institutions, parking garage booths, and executive reception areas. It is crucial that anyone who activates such devices, and those who are responsible for responding to such alarms, be trained to do so according to sound policies and procedures. Also, it is imperative that such devices be tested regularly.

Closed-Circuit Television

Closed-circuit television (CCTV) is a cost-effective security feature to help protect high-rise structures and their occupants. As one source notes:

> The primary purpose of a CCTV system is to enhance existing security measures and amplify the range of observation of security staff. To improve security, it may be useful to interface the CCTV system with intercoms and intrusion detection devices such as magnetic door contacts, motion detectors, or video motion detectors. The primary purpose of the video recording system is to record the picture from a camera and provide a permanent record for possible later review. As with other security systems and equipment, such as physical barriers, locking systems, and lighting and intrusion detection systems, CCTV is part of the basic security measures that make up the total security program. (Craighead, 2003, p. 153)

Another chapter of this manual provides a complete discussion of CCTV technology and its application to security. This chapter focuses on camera locations and their monitoring and response capabilities.

Camera Locations. The following locations are recommended for CCTV cameras in high-rise structures:

1. *Critical entry or exit areas for pedestrians.* These include the ground level exits from building stairwells, reception points on critical floors, mechanical floors, and crossover floors between elevator banks. The purpose is to monitor activity and, in the case of the fire stairs, to identify persons who leave the building via the stairwell enclosure.

2. *Pedestrian access points operated remotely.* These include pedestrian doors controlled from a remote location. The camera is needed to help ensure that entry is permitted only to authorized persons, without tailgating or piggybacking.

3. *Access points at which automated access control devices, such as electronic card readers, are featured.* The purpose is to help ensure that only those persons using the card are admitted and that tailgating does not occur.

4. *Sensitive interior spaces.* These are any spaces where remote surveillance provides a significant advantage. Such areas include mechanical areas, HVAC air-intakes, vaults, storage areas, elevator lobbies, and the interior of elevator cars.

5. *Vehicle entrances and exits.* Cameras at entrance and exit points can be used to monitor vehicular activity. For entrances where vehicle access is remotely controlled, a camera can help in determining whether the vehicle should be granted entry. For under-building garages, it is important to record images of the vehicles, including license plates, and the drivers who enter and exit.

6. *Covert surveillance areas.* Cameras can be concealed in a variety of ways and used to monitor and record activity in sensitive locations.[39] However, before they are used, local laws should be checked regarding their legality. Such cameras can be extremely useful in investigations, particularly of illegal activities such as theft.

When CCTV cameras are used in conjunction with remote access control, there is also a need for two-way voice communications between persons at the entry and control points.

Monitoring and Response. CCTV surveillance without critical analysis of the activity under observation, or without a response resource if one is needed, is operationally inadequate and economically wasteful.

Surveillance monitoring requires all of the following:

- output from each camera to be displayed at all times unless it can be alarm switched to activate the display when needed
- arrangement of the monitor screens to permit rapid visual analysis by a trained observer
- a means for bringing any particular camera of immediate interest to a monitor screen in the direct field of view of the operator
- the capability for permanently recording the display from one or more cameras, as needed, and for identifying the recorded material by camera location and clock time

In small systems with two or three cameras, the requirements just noted can be met easily with one monitor for each camera, plus a video recorder. For more elaborate systems, particularly those involving 10 or more cameras, a more sophisticated design effort is needed. In

---

[39] Of course, it is critical that such covert video surveillance not occur in areas where there is an expectation of privacy (such as restrooms, toilets, locker rooms, and changing rooms).

designing a CCTV system for a high-rise structure, it is important to remember this advice: "Design the application first and fit the equipment to it."[40]

Integrated Security Systems

It is important to have a central location within the high-rise structure where the security systems and equipment can be monitored and controlled. Another chapter of this manual discusses the design and operation of integrated systems (which by definition have a common control of a variety of sensors and field devices).

If a major integrated system is established, ideally no attempt should be made to use the system control center in any way other than as a remote facility. However, due to security staffing and cost considerations, sometimes the control area in a high-rise structure is established in the main lobby or near regularly used entry points. Monitoring personnel are expected to handle both the system demands and the spot demands of persons passing the area. The theoretical savings achieved by requiring that operators attend to other fixed-post duties may cause inattention to system-monitored events of far greater cost consequence. Such an arrangement also exposes the system control to compromise or attack. If the system has been properly designed, ideally the monitoring and control station should be the most secure location in the structure.

Security Personnel

Security personnel are an important aspect of the life safety and security of high-rise structures. Their primary role is to implement the building's life safety and security program. The number of security personnel necessary depends on the objectives of the security and life safety program and the role that security plays in it:

> The required levels of staffing for most high-rise office buildings are higher during normal business hours than after hours. During normal business hours, most tenants are open for business and there is an increased population of tenant employees, visitors, salespersons, tradespeople, building management staff, building contractors, couriers, delivery persons, solicitors, building inspectors, and others who may require the attention of building security staff. After hours, pedestrian (and vehicle) traffic usually lessens and only the number of janitorial staff increases (unless special activities are occurring in the building). (Craighead, 2003, p. 267)

Policies and Procedures

Clearly defined policies and procedures are an important life safety and security consideration for high-rise structures. Documented security instructions (commonly called post orders or

---

[40] Quote of Charlie R. Pierce, CPP, in Gips (2001, p. 82).

standard operating procedures) are an essential reference for security personnel to know what is expected of them.

3.4. SUMMARY

Generally, a high-rise structure is considered to be one that extends higher than the maximum reach of available fire-fighting equipment. Such structures have unique life safety and security needs. Since they often contain high concentrations of people and property, they need to be protected against threats that range from fires and natural disasters to criminal activity, including theft of assets, workplace violence, and acts of terrorism.

When certain types of incidents—particularly fires and explosions—occur, it is essential to immediately evacuate occupants to safety. However, due to the limited capacity of building emergency exit stairwells and elevators, not all occupants can leave the facility at once. Therefore, buildings must have up-to-date, thorough, and well-thought-out emergency management plans, and all occupants must be trained to react appropriately if an emergency should occur.

Many security and life safety features are available to help protect high-rise structures. They can be deployed in a manner that not only meets stringent building and life safety codes but also addresses security needs, which vary from structure to structure, depending on its type of tenancy and pattern of use. Sometimes life safety and security considerations appear to conflict with each other; therefore, special consideration must to be given to adequately address life safety requirements without compromising security objectives.

REFERENCES

9/11 Commission report: Final report of the National Commission on Terrorist Attacks Upon the United States (2004). New York: W. W. Norton & Company.

ASIS International Commission on Guidelines (2004). *General security risk assessment guideline*. Alexandria, Virginia: ASIS International.

ASIS International Commission on Guidelines (2004). *General security risk assessment guideline*, [Online]. Available: http://www.asisonline.org/guidelines/guidelinesgsra.pdf [2006, January 15].

ASIS International Commission on Guidelines (2004). *Threat advisory system response guideline*. Alexandria, Virginia: ASIS International.

ASIS International Commission on Guidelines (2004). *Threat advisory system response guideline*, [Online]. Available: http://www.asisonline.org/guidelines/guidelinesthreat.pdf [2006, January 15].

Black's law dictionary, 6th ed. (1990). St. Paul, Minnesota: West Publishing.

Clarke, F. B., PhD (2003). Fire hazards of materials. In A. E. Cote (Ed.), *Fire protection handbook*, 19th ed. Quincy, Massachusetts: National Fire Protection Association.

Craighead, G., CPP (2003). *High-rise security and fire life safety*, 2nd ed. Woburn, Massachusetts: Butterworth-Heinemann.

Donoghue, E. A., CPCA (2003). Building transportation systems. In A. E. Cote (Ed.), *Fire protection handbook*, 19th ed. Quincy, Massachusetts: National Fire Protection Association.

Edgar, J. M., & McInerney, W. D., AHC, CPP (2004). The use of locks in physical crime prevention. In L. J. Fennelly (Ed.), *Handbook of loss prevention and crime prevention*, 4th ed. Stoneham, Massachusetts: Elsevier Butterworth Heinemann. Used with permission of the National Crime Prevention Institute, School of Justice Administration, University of Louisville.

Emergency evacuation elevator systems guideline (2004). Chicago: Council on Tall Buildings and Urban Habitat.

Finneran, E. D. (1981). *Security supervision: A handbook for supervisors and managers*. Stoneham, Massachusetts: Butterworths, 1981. Quoted in L. J. Fennelly (Ed.), *Handbook of loss prevention and crime prevention*, 4th ed. Stoneham, Massachusetts: Elsevier Butterworth Heinemann.

Geringer, R. G. (1991, June). High-rises look to lock out problems. *Access Control*, p. 2.

Gips, M. A. (2001, April). Let's get digital. *Security Management*, p. 82.

Guidance for protecting building environments from airborne chemical, biological, or radiological attacks (2002). Publication No. 2002-139. National Institute for Occupational Safety and Health, Department of Health and Human Services, Centers for Disease Control and Prevention, May 2002, www.cdc.gov/niosh.

Hall, J. R., Jr., & Cote, A. E., PE (2003). An overview of the fire problem and fire protection. In A. E. Cote (Ed.), *Fire protection handbook*, 19th ed. Quincy, Massachusetts: National Fire Protection Association.

Healy, R. J. (1983). *Design for security*, 2nd ed. New York: John Wiley & Sons.

Holmes, W. D., PE (2003). Occupancies in special structures and high-rise buildings. In A. E. Cote (Ed.), *Fire protection handbook*, 19th ed. Quincy, Massachusetts: National Fire Protection Association.

Kitteringham, G. W., CPP (2001). A study of two types of vertical crime pattern analysis in the commercial, multi-tenanted high rise structure. Thesis for Master of Science in Security and Crime Risk Management, Scarman Center for the Study of Public Order, Canada, February 2001.

Lathrop, J. K. (2003). Concepts of egress design. In A. E. Cote (Ed.), *Fire protection handbook*, 19th ed. Quincy, Massachusetts: National Fire Protection Association.

Milke, J. A., PhD, PE, & Klote, J. H., DSc, PE (2003). Smoke movement in buildings. In A. E. Cote (Ed.), *Fire protection handbook*, 19th ed. Quincy, Massachusetts: National Fire Protection Association.

National fire code (2001). Quincy, Massachusetts: National Fire Protection Association.

Puchovsky, M. T., PE (2003). Automatic sprinkler systems. In A. E. Cote (Ed.), *Fire protection handbook*, 19th ed. Quincy, Massachusetts: National Fire Protection Association.

Roberts, J. C., PE (2003). Automatic fire detectors. In A. E. Cote (Ed.), *Fire protection handbook*, 19th ed. Quincy, Massachusetts: National Fire Protection Association.

Schifiliti, R. P., PE (2003). Notification appliances. In A. E. Cote (Ed.), *Fire protection handbook*, 19th ed. Quincy, Massachusetts: National Fire Protection Association.

Shapiro, J. M. (2003). Standpipe and hose systems. In A. E. Cote (Ed.), *Fire protection handbook*, 19th ed. Quincy, Massachusetts: National Fire Protection Association.

Taylor, G. M. (2003). Halogenated agents and systems. In A. E. Cote (Ed.), *Fire protection handbook*, 19th ed. Quincy, Massachusetts: National Fire Protection Association.

Webb, W. A., PE (2003). Air-conditioning and ventilating systems. In A. E. Cote (Ed.), *Fire protection handbook*, 19th ed. Quincy, Massachusetts: National Fire Protection Association.

ADDITIONAL SOURCES

Broder, J. F. (2000). *Risk analysis and the security survey,* 2nd ed. Woburn, Massachusetts: Butterworth-Heinemann.

Emergency evacuation elevator systems guideline (2004). Chicago: Council on Tall Buildings and Urban Habitat.

Emergency planning guidebook (1994). Washington, DC: Building Owners and Managers Association.

Emergency planning handbook, 2nd ed. (2003). Alexandria, Virginia: ASIS International.

Hopf, P. S. (1979). *Handbook of building security planning and design.* New York: McGraw-Hill.

NFPA ready reference: Fire safety in high-rise buildings (2003). Quincy, Massachusetts: NFPA International.

San Luis, E., Tyska, L. A., & Fennelly, L. J. (1994). *Office and office building security,* 2nd ed. Woburn, Massachusetts: Butterworth-Heinemann.

U.S. Marshals Service (1995). *Vulnerability assessment of federal facilities.* Washington, DC: Government Printing Office.

World trade center building performance study: Data collection, preliminary observations, and recommendations, (2002). FEMA 403. Washington, DC: Federal Emergency Management Agency.

www.ctbuh.org. The Council on Tall Buildings and Urban Habitat (CTBUH) was established to facilitate professional exchanges among those involved in all aspects of the planning, design, and operation of tall buildings. CTBUH publishes a list of the 100 tallest buildings in the world.

www.emporis.com. This site contains a valuable database of high-rise buildings throughout the world.

PART 4
Integrated Security Systems Design and Specification

This text is excerpted from the March 2008 revision, published as "Part III: Integrated Security Systems Design and Specification" in the "Assessment" chapter of the *Protection of Assets (POA) Manual*.

4.1. INTRODUCTION

Proper use and application of the integrated security systems design process is the single most important element in the defense against dynamic threats and potential catastrophic losses. In other words, a fence used primarily to delay entry is only one of the four elements of physical design which are territorial definition, surveillance, building forms and compatible building placement. Typically, design and integration are performed to introduce and meld technological and physical elements into the overall asset protection program. When carefully and diligently followed, the process results in a fully integrated security program that blends architectural, technological, and operational elements into a flexible, responsive system. For example, a risk analysis of a data process operation may determine significant loss exposures.

Integrated security systems designs can address any number of security subsystems or elements to form a complete system. Particularly important factors in system design are the particular environment or unique needs of the facility requiring protection. Anticipated threats, risks, vulnerabilities, and constraints all need to be taken into consideration to best pinpoint the best solution. While physical barriers are either natural or structural only, integration is the logical, symbiotic combination of these elements into a system. Options or elements available to the designer include the following:

- facilities, architectural barriers, and spatial definition such as gates, physical barriers, and hardening to deter, delay, and deny an adversary

- access control and locking systems and procedures to establish concurrent levels of security control and to channel site and facility personnel and vehicular traffic and to control access to a site, its parking lots, facilities, and internal facility areas designated critical by the owner

- closed-circuit television (CCTV) systems to provide area surveillance, detection, assessment, and archived retrieval of views of external areas, internal facilities, and site vulnerabilities, and to assess risk-causing events

- aesthetically placed, deterrent-based electronic intrusion technologies designed to detect and in some cases assess and electronically react to intrusion and unauthorized access attempts into an owner's site or facility

- environmental design concepts and strategies to create smooth traffic patterns, instill defensible space, and prevent the incidence and fear of victimization on owner property

- communications networks and systems to collect, integrate, transmit, control, and display alarm and other data at the local or central monitoring location for notification and direction of alarm response and conduct of security operations

- comprehensive security policies and procedures to instill responsibility, inform personnel of security and safety issues, and provide security program control over designated assets residing at the center

- information-based decision support systems to ensure a proactive response to all forms of risk to persons, property, facilities, and operations

- intellectual property, proprietary information, and trade secret protection strategies, including information technology risk prevention and mitigation strategies

This document provides an overview of the tasks and players involved in a security systems implementation project from initial inception through project completion and systems operation. The basic tasks of security systems implementation are as follows:

- planning and assessment to determine security requirements

- developing conceptual solutions for resolving vulnerabilities

- preparing security systems design and construction documentation

- soliciting bids and conducting pricing and vendor negotiations

- installing, testing (which is most likely to be overlooked), and commissioning the security systems

System in the security context is a combination of equipment, personnel, and procedures, coordinated and designed to ensure optimum achievement of the system's stated objectives. A system includes more than hardware components. A protective system is evaluated on the cost effectiveness of individual measures in countering threats, reducing vulnerabilities and decreasing risk exposure. Although much of the following discussion is related to security technology, the process is also applicable to the design, procurement, and deployment of other security elements.

4.2. SYSTEMS DESIGN PROCESS

Figure 1 provides an overview of the systems design process.

In simpler form, the process starts with the planning and assessment phase. The first task of that phase is the identification of critical assets, potential threats, subsequent vulnerabilities, likely risks, and functional requirements. For example, in order to deter burglary, a common method is to harden the target to deter or reduce the opportunity to commit a crime. In a broader sense, it is the time to collect information on security needs, objectives, and constraints so that risk management and control can be effected before a security event rather than after. This concept of proactivity versus reactivity in security planning is a key aspect of effective risk management and control. The second task of the planning and assessment phase is to analyze security requirements and formulate solutions or countermeasures concepts to reduce

Figure 1
Systems Design Process

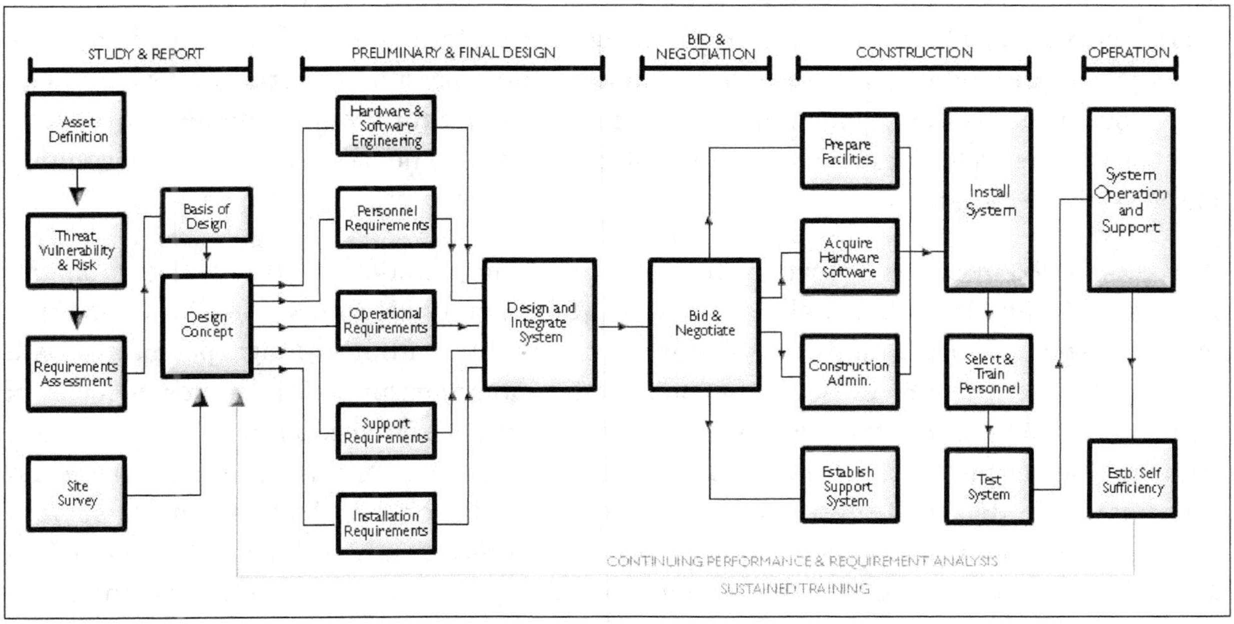

or eliminate vulnerabilities and mitigate risks. Once these concepts have been validated both operationally and in budgetary terms, the design phase can begin. The output of the design phase is systems hardware and software identification, placement, integration, and performance documentation that is sufficiently clear and complete to ensure consistent, accurate interpretation by suppliers and installers for systems procurement and implementation.

The systems design process is a serial process. Each phase and task must be performed sequentially before the next can begin—the output from one task becomes the input for the next. This is an important concept for senior management. Depending on the nature of the environment, organization, and potential risks, the process requires that a significant effort be expended to develop the basis of design and resultant design documentation and construction implementation. The process can be shortened only a little when the user arranges for a design/build relationship with a contractor versus a design, procurement, and construction relationship with an architect or owner. Generally, increasing staff or budget cannot substantially shorten the process. Moreover, if the security systems project is part of a major facilities construction or upgrade project, security systems design and implementation will be executed and documented according to the schedules and documentation of architectural design and construction, normally managed by a certified architect as project overseer. In that case, those responsible for the security design will work directly with the architect and his or her project team to document security requirements and include them in contract documents for bid and construction.

4.3. PLANNING AND ASSESSMENT PHASE

The planning and assessment phase is the initial phase of any security design project. This phase consists of gathering all relevant pre-design asset information and analyzing it in terms of project requirements and constraints. Planning and assessment efforts always culminate in a security "basis of design," the first and most important output of the design process. The basis of design focuses on specific project requirements and a conceptual design solution based on those requirements.

A security problem can be defined by three loss event considerations such as profile, probability and criticality. It is almost always an asset vulnerability identified during the planning and assessment process, and the solution to this problem from a design perspective is a security countermeasure or series of integrated countermeasures. The principal focus of security planning and assessment is on assets that require protection, and the protection strategy that results is based on threats, vulnerabilities, risks, and requirements based on those assets to determine if the impact is high, medium or low. Assessments or surveys provide demographic data, such as age, sex, education level and race of the member population as well as a determination is a plan is functional and current. An important component is a checklist so that later an operational audit can build on the original assessment or survey. Any other approach will miss the mark and result in inadequate or improper security.

Thus, security planning, assessments and operational audits are formal processes for identifying and analyzing the security issues and problems associated with asset protection, and of developing asset protection requirements, objectives, criteria, concepts, and methods that will be used in the eventual detailed design of the solution. The assessment or survey is effective if action is taken on the recommendations and results are measured against acceptable standards. This phase involves considerable teamwork between operational, facilities, engineering, and architectural representatives to present the proposed solution to management for approval and budgeting.

Three key ingredients in the planning phase determine its eventual success. First, a multidiscipline and committed approach from either a single individual or project team is fundamental to success. Second, spending the necessary time and effort in the planning phase ultimately results in a more accurate and responsive design solution, reduced risks, reduced overall costs of potential losses, and increased longevity and effectiveness of the installed systems. Third, decisions made during the planning and assessment phase must be made on the basis of sound and relevant risk and asset environmental information. In essence, security design is just as dependent on collecting good data leading to informed decisions by knowledgeable people as is any other analytic process where a solution is engineered and constructed.

The outcome of the overall planning phase is a set of security requirements, or objectives that are used as a basis of the eventual design (also called design basis). Assessment consists of surveying and analyzing the assets and protection, normally through an initial site survey, and applying the risk assessment and design process to arrive at a conceptual solution based on derived protection requirements. Thus, the planning and assessment phase results in a conceptual design solution that categorizes vulnerabilities by their criticality and identifies the most preferred and cost-effective protection scheme to mitigate or eliminate asset risks. The initial design solution at this phase of the process is entirely based on the designer's interpretation of functional requirements in the conceptual solution. Without these requirements, there can be no meaningful design solution and precious capital can be wasted in eventual expensive construction.

Another important outcome of the planning phase is the development of the business case for the new or upgraded security systems because systems will not only be evaluated on quality and reliability, but also cost. The business case documents the impact of the design solution on the business, the necessary investments, expected quantifiable savings, and other metrics that allow decision makers to make investment decisions on a security project. The main feature of a security business case is a series of economic metrics (return on investment, payback, net present value of cash flow, etc) that are used to justify the security solution up the management chain. A formal presentation on the security needs, business case, costs and benefits, alternatives, and impact on operations is often mandatory before the expenditure of capital. In the architectural and engineering world, the planning phase is typically referred to as the programming and schematic design phase leading to a design basis (requirements analysis) and conceptual design.

4.3.1 REQUIREMENTS ANALYSIS

Effective planning requires a requirements analysis. Requirements analysis involves an assessment of risks and leads to protection solution requirements. Most security professionals use the terms, risk, threat and vulnerability interchangeably, however below the terms are further defined. The following discussion outlines elements of requirements analysis.

Assets

The first and most important factor is an identification of the organization's critical assets—those that are essential to the well-being, profitability, and continued operation of the organization. In a security context, assets include employees and their productivity, the corporate image, production and distribution operations and processes, information, and intellectual property—in addition to plant, equipment, negotiable documents, cash, inventory, parts, and finished goods. A thorough evaluation entails a quantitative valuation of costs associated with the replacement of lost or damaged assets. The true cost should consider the loss of production, productivity, and market position. Assets should be categorized according to whether they are critical or merely supporting.

Threats

The second factor in requirements analysis is the determination of possible threats against each of the identified assets. Threats are the sources or methods of harm and are categorized as criminal, natural and accidental. Threats to an organization are always present, but the selection of a particular organization for attack depends on many variables. Today, access to an organization's critical assets—including its employees, information base, products, and network communication channels—can lead to a serious and sometimes catastrophic disruption of business operations and a devastating loss of market position and image. Even the traditional types of workplace crimes and incidents can seriously disrupt a business by creating unsafe and insecure feelings in employees and customers alike.

While threats cannot always be eliminated, anticipating them can lead to informed placement of safeguards. For security assessment purposes, one must consider the full range of conventional and unconventional threats to the safety and security of an organizations' employees,

INTEGRATED SECURITY SYSTEMS DESIGN AND SPECIFICATION

business, and facility. Proper planning relating to threat assessment will result in the determination of the degree of security required in all areas of the company. The following is a summary of adversary categories and capabilities that must be assessed during the planning phase:

- **Unsophisticated criminals:** those with no skill or technique in undertaking attacks
- **Sophisticated criminals:** skilled aggressors highly trained to target and disrupt facilities and destroy or steal products
- **Disgruntled employees or contactors:** employees or contractors who are discontented—harm could include revenge by an angry employee; employee-to-employee assaults; theft of information, trade secrets, products, supplies, or equipment; theft of other employees' property; and equipment sabotage
- **Disoriented persons:** people who are confused or unsettled and take their problems out on the organization or an employee
- **Activists/protesters/extremists:** people who object to the company's activities or success and conduct an attack on facilities or personnel
- **Terrorists/subversives:** those who attack civilians and other targets, such as facilities or products in transit, for political, religious, or similar motives

Threats may be historic (having happened to this organization, in this geographic area, in this industry, etc.) or perceived. Threat information can be gained from police reports, crime statistics, incident reports, neighboring or like organizations, and many other sources. Threat sources are then categorized according to their applicability to specific assets and their likelihood.

Vulnerabilities

Vulnerabilities are the perceived exposures created when a potential threat source targets a specific asset for attack, taking into consideration the organization's existing asset protection scheme. Vulnerabilities are weaknesses in an organization's asset protection plan; can be identified in an assessment and addressed in a security design. The first phase of the design and evaluation process is to determine the system's objectives.

Vulnerabilities are identified through surveys using vulnerability self-assessment tools (VAST), plan reviews, and discussion of proposed project designs which makes the task of risk analysis more manageable. By establishing a base from which to proceed a designer may identify specific vulnerabilities that need to be better protected after considering the critical assets, threat sources, and likelihood of attack. When feasible, it is always beneficial to complete the risk analysis during the design phase of the project.

Functional design requirements are almost exclusively based on the vulnerabilities of specific assets. However, these requirements are not prepared until the formal risk analysis has been completed. While the main purpose of vulnerability analysis is to identify existing asset exposures, the designer also uses vulnerability information to determine the adequacy of existing protective measures and to assess the need for and extent of additional protective measures in the design.

Risks

Risk, the possibility of an undesirable event, is a quantitative measure of the likelihood that an identified threat source will target specific assets and cause harm or loss. The likelihood that a risk will affect the loss of assets is known as loss event probability. A loss event profile is undertaken to determine the kind of risks that might affect the assets. Three risk factors used to identify risk exposures are types of risks, probability of occurrence and loss potential. From a design standpoint, there are two key aspects of risk. First, the designer seeks to minimize each identified risk of harm or loss through the application of security strategies or measures in the design. In other words, a primary question to ask is why are we doing this countermeasure? If the designer determines that existing measures are sufficient to deter an adversary from attack, he or she will include those in the design. Second, while the designer evaluates and quantifies risks, and then prioritizes the requirements for security measures based on the criticality and vulnerability of the assets, he or she may designate certain risks to management for a determination of whether to accept (level of loss acceptable without severely disrupting operations), transfer (amount of loss through disappearance or destruction that could be tied to an occurrence in an insurable amount), or avoid the risks (amount of loss that likely would occur without the security program) if the necessary countermeasures might not be cost-effective or appropriate.

Risk analysis makes it possible to determine the scope and potential annual costs of loss expectancy due to impact and frequency so an organization can apply the right level of protection. For example, one could research crime records to establish frequency and types. Generally the least expensive countermeasures are procedural controls. Risk analysis is the key phase of requirements analysis and leads to protection requirements, not specific solutions.

Risk is also associated with the design and layout of the facility and its existing security measures. Dynamic risk carries the potential for both benefit and cost reduction. A site survey and security audit, or a study of plans for a new building, will assist in quantifying risk at current security levels. The security survey needs to take into consideration the natural threats not only due to the location of the facility, but also the psychological factors that may detract, or contribute to the overall protective scheme. By doing so to the appropriate resources and security measures are channeled to adequately protect against the most probable threats.

INTEGRATED SECURITY SYSTEMS DESIGN AND SPECIFICATION

4.3.2 DESIGN REQUIREMENTS

The next task, defining design requirements, is used to develop specific functional design guidance leading to security strategies. Before looking at specific asset protection requirements, it is useful to formulate a statement of the overall objectives or mission of the integrated security system (ISS). The objectives must reflect and support the overall corporate mission if the ISS is to be funded and supported by management. The overall protection objectives should be validated as each design planning task is completed. New insights will come as the process develops. It may become necessary to revise the mission statement, but at the end of the task the requirements definitions should accurately reflect the overall asset protection objectives.

The designer develops security requirements by examining each vulnerability and defining what protection functions are required to mitigate, reduce, limit, or erase it. For example, if terminated employees can enter the facility freely to harass or assault management or other employees, the security requirement may specify a means to detect terminated employees and delay their access. This requirement may be further expanded to address all unauthorized entries, a proper response protocol, and a method of recording. In this example, technology, security staff, and procedures are integrated into the requirement.

It is useful to add a level-of-confidence factor to each functional security requirement. The preceding paragraph uses the terms *detect* and *delay* rather than *prevent*. A security system built on absolute objectives, such as total denial of unauthorized entry (100 percent confidence), will either be impossible to design or so costly as to be impractical. An unauthorized entry is not the real danger, per se, but rather the consequences of such entry. Thus, the requirements definition should focus on preventing, delaying, or modifying the consequences.

Design solutions to various asset vulnerabilities may be the same, similar, or complementary. For example, a security requirement that leads to the detection and delay of unauthorized access may be partially or completely applicable to another perceived asset vulnerability, such as the prevention and deterrence of trade secret theft by a business competitor. A thorough planning process must evaluate all asset vulnerabilities and list specific functional requirements and resultant protection strategies.

One calculation tool that can be used for detection and interdiction is called the EASI Model (estimated of adversary sequence interruption) which is a simple tool that uses detection, delay, response and communication values to determine what are the effects are on these values when the physical security parameters are changed. As with any security system, timely detection is key to allow adequate time for a response and interdiction of an adversary. (Additional information on EASI can be found in Garcia's *The Design and Evaluation of Physical Protection Systems, 2nd Edition*.)

The level of protection for a group of assets must meet the protection needs of the most critical asset in the group. To have a balanced protection system, no one element can be more quickly penetrated than another. However, the designer of a security system may separate a critical asset for specific protection instead of protecting the entire group at that higher level. Thus, the requirements analysis and definition process is designed to do the following:

- Ensure that the selected solutions will mitigate real and specific vulnerabilities.
- Provide a cost/benefit justification for each solution.
- Identify all elements (technology, staffing, and procedures) and resources required for each solution.
- Provide a basis for the accurate and complete system specification that will be used to procure and implement the solutions.

Figure 2 provides an example of a requirements analysis completed for a specific project.

Figure 2
Requirements Analysis

The entrance to the building is at the first floor, and is flanked by a Real Estate office and a Bank. There are two entrances, a north entrance and a south entrance. There is an exit door at the east side of the building. The Real Estate office and Bank will have staff members and visitors accessing its facilities. There must therefore be a demarcation of facilities on entry to the building. Security is essential at this level as the elevator lobby and Day Care center are susceptible to entry by visitors other than those coming to visit.

Floor	Door	Use	Assets	Criticality	Threats	Vulnerabilities	Risk
First	1	Access to terrace from within the office	• HR Office • Personnel files	Med	• Break-in • Arson • Theft of records • Theft of equip • Door left open	• Fire engulfing whole floor • HR info being compromised	• Injury to personnel • Loss of records • Disruption to work processes • Psychological danger
	2	Secondary door to day care facility	• Day care center equipment • Children records • Children	Med	• Break-in • Theft of records • Theft of equip • Door left open • Child abduction	• Compromise of safety and care at Day Care Center	• Injury to personnel and children • Loss of records • Disruption to work • Psychological danger

4.3.3 BASIS OF DESIGN

Once the requirements definition is complete and the individual design requirements are identified, the designer prepares a basis of design and submits it to the design team. The basis of design documents the initial designation of assets deemed critical, outlines the overall objectives of the asset protection program, describes the results of the risk analysis, lists the functional requirements to be satisfied by the eventual design, and provides a narrative operational description of the proposed systems, personnel, and procedures that constitute the security system or program.

The basis of design becomes the designer's means to obtain consensus from the design team on the goals and objectives of the project, what will constitute the project, and how the project will secure the assets. Each member of the design team should have input into the design basis. This is not the time to identify engineering details, prepare budgets, or identify and debate specific countermeasures. This is the time when the project is first conceived, the requirements are derived from a rigorous risk assessment, and subsystem functional descriptions are provided to indicate eventual system performance. Also, this is the time when initial site surveys may be accomplished to gather information on existing conditions and measures and on any needs for upgrades or additions.

4.3.4 CONCEPTUAL DESIGN

The conceptual design, also called a design concept, is the last task of the planning and assessment process. In this task, the designer formulates a complete security solution for the assets to be protected. The security solution typically consists of protection strategies grouped together to form an asset protection program or augment an existing one. Thus, the solution normally includes security systems complemented by procedures and personnel. At the conceptual stage, however, the solution is expressed in general narrative and descriptive terms accompanied by an initial budgetary estimate for design and construction.

The design concept incorporates the basis of design; documents the findings, conclusions, and recommendations from any initial surveys; and is the first opportunity to document the project's design. From an architectural perspective, the design concept is usually referred to as the initial conceptual design or schematic phase.

The security designer collaborates with the site owner on an integrated, holistic approach to asset protection. The designer is principally concerned with establishing protective measures generally configured in concentric rings around protected assets to make it progressively more difficult for an intruder to reach critical targets and escape undetected. These protection-in-depth or redundant schemes build barriers or time delays into the intruder's path to protected assets and make it possible for other security resources to respond.

The importance of having a redundant security system is based on the ten principles of probability developed by the French mathematician and astronomer Marquis de Laplace (1749-1827). According to Laplace formula, when events are independent of each other, the probability of their simultaneous occurrence is the product of their separate probabilities. Meaning that the probability of one detection system in the security system being circumvented is high, but the probability of all the detectors and barriers in an in-depth or redundant security scheme being compromised is very low.

There is some debate as to the level of detail included in a design concept. Normally, the concept includes the elements noted above, plus some initial design detail. A design concept's detail should never be more than a top-level description of the various anticipated security system elements, subsystems, and support systems.

The intended subsystems should be narratively described in the concept, as should their interaction with one another to form a complete system. The narrative details should be accompanied by representative details on specific, anticipated design aspects, such as an access controlled entry door, an emergency exit, etc. Finally, overall block diagrams should be prepared depicting systems, subsystems, and representative element-level connectivity accompanied by project construction cost estimates.

For architectural purposes, it is common to mark up architectural floor plans with intended devices, control points, and system connectivity to give the project planners an indication of the scope and depth of interface (power, etc.) for the security portion of the project. Architects thrive on detail and take every opportunity to demand that that level of detail be included early in their project leaving.

At some point, the project must be approved. The concept level is the ideal time to seek management approval since the project team has reached consensus on the project's scope and sufficient detail has been developed to create an initial budget.

The designer's choice of countermeasures depends largely on their cost-effectiveness. Cost-effectiveness criteria that might be used include operational restrictions, nuisance alarm susceptibility, installation cost, maintenance costs, probability of detection, mean-time-between-failure, contribution to manpower reduction, contribution to asset vulnerability, and reduced risk, normally expressed in the monetary consequence of loss or destruction.

A designer may choose from many countermeasure options. Most security designers identify four principal security strategies—prevention, detection, control, and intervention—as the most important functional requirements of security design. Homeland security features five principal strategies: preparation, prevention, detection, response, and recovery. Figure 3 shows a sample countermeasures development table.

INTEGRATED SECURITY SYSTEMS DESIGN AND SPECIFICATION

Figure 3
Countermeasures Development Table

FLOOR	DOOR	USE	ASSETS	CRITICALITY	COUNTERMEASURES
Basement	1	Loading Dock	• Incoming Materials	Med	• Surveillance • Access control
	2	Shipping/Receiving/Storage	• Stocks/Products	Med	• Access control
	3 & 4	Door leading into corridor	• General equipment	Low	• Intercom • Access control
	5 & 6	Shipping/Receiving office	• General equipment • Stocks	Med	• Duress • Door release • Monitor

4.3.5 DESIGN CRITERIA

Design criteria constitute the ground rules and guidelines for the design. In effect, these are additional design requirements that the designed must consider along with risks. The criteria fall into a number of categories, some based on expected system performance, some on operational and financial considerations, and others on style, design, codes, and standards. Not all asset protection measures are possible or practical. Other criteria will identify constraints or limitations that apply to the design, implementation, and operation of the system.

At this phase of the design process, it might only be necessary to list the criteria rather than include a complete description of the details. The details will be included in the design specifications and construction or contract documents. Some of the more influential design criteria will include the following:

Codes and Standards

Particularly for facility security design and upgrade projects, design and implementation will probably have to follow national and local building, fire, and life safety codes. Applicable codes must be identified and applied to the initial design to ensure compliance. In addition, various laws may come into play, including those regarding security officer registration and training. Also, the organization may have its own set of standards for design, procurement, modification, and construction, such as work rules, insurance coverage, acceptable color schemes, and competitive bidding rules. Some organizations even have a set of security standards or guidelines that establish design and construction standards for security system implementation. Certain life safety codes have a significant effect on the selection, configuration, and operation of components selected to control doors. Failure to adhere to these codes and standards may lead to eventual rejection of the design solution in the construction phase, and meeting codes may require expensive changes to the constructed system. This occurs particularly where security controls are applied at junctures in the building's established path and it is later determined that such controls violate the safety code and must be eliminated. Other cases involve the use of particular locking mechanisms (such as electric strike, electro-magnetic, and vertical pin locks), their application to certain door types, and the resultant door hardware configuration necessary to meet codes.

Quality

A designer should always be aware of the quality and performance differences between components. Generally, the use of quality components in a superior design goes a long way. A good design always strikes a balance between quality components and overall cost. Quality also needs to be applied consistently. For example, it makes little sense to install a high-quality lock in a hollow wooden door or a metal door surrounded by simple drywall construction or to install an intrusion alarm system with sensors and a control unit but without an annunciator. The alarm system must also have tamper protection that provides an alarm signal if the system is compromised. The designer always identifies options to make it easier for management to understand cost drivers and the relative performance of different configurations. It is also important to document the trade-offs between cost and quality.

Capacity

Capacity, size, and space requirements are major determinants of security system solutions. Desired capacity (for example, number of card holders for an access control system, number of alarm zones monitored, number of access controlled doors, etc.) may be changed as the design is developed. Still, having a general estimate at this stage reduces the number of design iterations. Nothing complicates a design more than a restatement of system capacity requirements midway through the design process. The designer always considers expansion capacity in the design from the very beginning, typically adding anywhere from 10 to 15 percent spare capacity.

Performance

Component performance is usually detailed in a performance or project specification. Overall system performance parameters, however, should be stated as design criteria in the design basis documentation as well, especially if the designer intends for systems to interact with existing systems or conditions. The following are examples of performance parameters:

- The access control system must connect to an existing local area network.

- The access control system must effectively manage personnel traffic at shift changes.

- The card reader-controlled turnstile subsystem must have a minimum throughput of 500 badge holders per hour and be able to accommodate building evacuation within 10 minutes.

The performance list can also include reliability and maintainability criteria—for example, that the turnstile must have a mean time between failure (MTBF) of 2,000 hours.

Features

Major system features should be summarily defined in the basis of design documentation and eventually in more detailed terms in the performance specification. A good example is the placement of optical turnstiles in the lobby of a high-rise building, based on throughput and evacuation requirements. The throughput feature usually dictates the number of lanes, and most lobbies can accommodate only so many lanes. If the design basis requires functions that require design features that are not commonly available, procurement competition will be limited and costs could escalate. Custom features may also complicate component interface, require additional procurement and implementation time, and be more difficult to maintain. Designers should have a detailed knowledge of performance features that are normally available off-the-shelf. It is worthwhile to perform a reality check of both systems performance and feature design criteria with several manufacturers before finalizing the list.

Cost

Two of the main cost drivers for security design are the design fees and projected system construction costs. Regarding design costs, some owners elicit the assistance of installer/integrators to design systems, thereby saving design costs. Others prefer to seek professional assistance from a knowledgeable consulting engineer. Over the long haul, it is beneficial to have a knowledgeable person lead the integrated design process. That person's knowledge can help reduce costs of construction, personnel, and procedures. Some people experience shock when they see how expensive reasonable security can be, particularly integrated security systems involving access control, intrusion detection, and CCTV. If the risk analysis has been thoroughly documented and is quantitatively based, then additional funding may be easier to justify. A budget is often a required design goal and should be included as one of the initial design criteria.

Operations

Two main criteria drive security designs. First, security programs need to have minimum negative impact on productivity and facility operations. Restricted access in production areas may affect operations, especially if those areas experience high volumes of traffic. Hence, operations managers should be consulted early in the process to find alternate solutions (such as a new layout for a production area). Second, security operations should be seen as a natural use of security systems. For example, a systems design should include the capability to adjust to both shift changes and normal patrol operations. A good system design allows for timing of system activations while also providing for a central location where alarms, video surveillance, and communications can monitored.

Culture and Image

Corporate culture is a significant factor in the design and implementation of security systems and programs. Culture is what distinguishes one organization from another, and it determines how security is defined and implemented in a particular organization. Care must be taken to ensure that procedures and training maximize people's acceptance of change. Related to culture is image, the perception of the organization by the outside world. Several factors, such as customer service, promotional activities, and exterior and interior facility design, help to form an image. If the security function is to support corporate goals, the security program must reflect the corporate image. Whether the program emphasizes high-profile or low-profile security, it should always consider the aesthetics of visible security components, such as security officer uniforms or security equipment. Some of these design topics may be covered by the corporate standards discussed earlier; criteria not covered above should be listed here.

Monitoring and Response

An essential component in any security system and program is the design of a centrally located security operations center and the assignment of security staff to monitor alarm systems and respond to alarm conditions. The design of a central monitoring facility is becoming more important as the need for business continuity increases. In addition, as more organizations apply integrated security systems on a global basis, effective and efficient monitoring and response may become even more important. If the budget restricts the design of a security operations center or the availability of suitable staff, the system design will need to minimize monitoring requirements or personnel or incorporate the capability for outside monitoring. For enterprise systems, a third-party monitoring and response arrangement calls into question the investment in an enterprise system in the first place. However, for some remote or single locations, intrusion detection and access control alarms may report to a commercial central alarm station or be annunciated and controlled through a proprietary, on-site system. CCTV systems can also be applied on an enterprise basis. Such systems can be remotely monitored and used as assessment tools for local alarms, and they can also be used as archival mechanisms to retrieve previous alarm or transaction scenes at selected points across the enterprise network. The monitoring and response function used in rudimentary systems for small or medium-sized facilities can be provided by a central station. For more complex systems and larger organizations, the preferred method of monitoring and response is by on-site security staff in a properly designed and outfitted security operations center. In some cases, organizations choose to use on-site staff during the business day and remote monitoring after hours. For on-site functions, the skills and training of staff should match the complexity of the monitoring, control, and response systems.

Preliminary Cost Estimate

The last major task in the planning phase is to develop an initial budget, both for capital expenditures and recurring costs associated with the proposed system. Since at this early phase no detailed design work has been performed, nor have component quantities been finalized, the budget can be a conceptual, order-of-magnitude estimate at best. Some designers with experience in estimating the systems to be implemented can estimate within 10 percent of final bid prices. Most people, however, need assistance from vendors, manufacturers, or contractors to obtain MSRP (manufacturer's suggested retail price) for the equipment, installation, software, and support systems. The services of a knowledgeable, independent security consultant may be required.

One danger of an inadequate initial budget is that the designer may have to repeat a lengthy, difficult budget approval process. If a specific hardware vendor will supply most of the equipment desired, that vendor's expertise can be helpful in developing an initial estimate. Although that estimate is conceptual, its accuracy is important. Generally, it should be within 15 to 20 percent of final bid prices. If it is too low, later discovery of the real cost of the project could lead to its cancellation or insufficient funding to construct a totally responsive security system. If the cost estimate is too high, the initial budget may not be justifiable and may not be approved.

The following are examples of items that should be considered in the estimate:

- Capital projects
    - all equipment and support systems and their installation cost, including all primary and backup systems, software, components, mounting hardware, sensors, termination panels, control panels, back boxes, junction boxes, conduit, cable, battery backup power, uninterruptible power supplies, and main power circuits
    - freight, taxes, etc.
    - project management and supervision labor
    - shop drawing submissions
    - testing
    - commissioning
    - operator and user training
    - as-built (record) drawings
    - warranty
    - design fees

- Service projects and recurring costs
    - security staff payroll, including supervision, benefits, holidays, vacations, and sick leave
    - uniforms and equipment
    - training
    - equipment maintenance, repair, and replacement
    - consumable supplies, including printer paper, ink and toner cartridges, and backup media
    - replacement access control cards, badges, and review of development procedures
    - central alarm station monitoring and response

Figure 4 shows one type of detailed cost estimate format used for design planning purposes. For early phases of the project, the estimate would provide a single, lump-sum estimate for

Figure 4
Sample Cost Estimate Format (Circa 2007)

FACILITY SECURITY SYSTEM COST ESTIMATE										
Head End	Manufacturer	Model	Unit Cost	Quantity	Base Price	Installation Time	Hours Labor	Labor Cost	Installed Cost	
PC Workstations (Computer/Monitor/Keyboard)	Dell	Optiplex GX1p	$2,800.00		$0.00	4	0	$0.00	$0.00	
Application Software – Network	WSE	NSM	$7,500.00		$0.00	8	0	$0.00	$0.00	
Alarm/Access Printer	Epson	570-PRT	$600.00		$0.00	1	0	$0.00	$0.00	
Network Hub	Lancast	4490	$285.00		$0.00	1	0	$0.00	$0.00	
										$0.00
Access Control										
Access Control Panel	WSE	4100	$2,445.00		$0.00	6	0	$0.00	$0.00	
Alarm Input/Output Board	WSE	MIRO 16/8	$1,090.00		$0.00	1	0	$0.00	$0.00	
Signal Multiplexer	WSE	Nexstar	$300.00		$0.00	1	0	$0.00	$0.00	
Enclosure (30" x 36")	WSE	92410080000	$250.00		$0.00	1	0	$0.00	$0.00	
Panel Power Supply	Alarm Saf	PS-1	$500.00		$0.00	1	0	$0.00	$0.00	
Lock Power Supply	Alarm Saf	PS-5	$280.00		$0.00	1	0	$0.00	$0.00	
Proximity Cards – Thin	WSE	QuadraKey	$5.00		$0.00	0	0	$0.00	$0.00	
Card Reader – Proximity (Surface Mount)	WSE	DR4205	$480.00		$0.00	1	0	$0.00	$0.00	
Network Connection			$195.00		$0.00	1	0	$0.00	$0.00	
Modem/Telephone Connection			$195.00		$0.00	1	0	$0.00	$0.00	
Terminal Server	WSE	Cobox E2	$545.00		$0.00	1	0	$0.00	$0.00	
										$0.00
Photo ID Equipment										
Photo Identification System PC	Dell	Optiplex GX1p	$2,800.00		$0.00	4	0	$0.00	$0.00	
Photo Identification System Software	WSE	QuikWorks 4	$4,250.00		$0.00	2	0	$0.00	$0.00	
Photo Identification System Camera	Kodak	DC210	$900.00		$0.00	1	0	$0.00	$0.00	
Photo Identification System Frame Grabber	WSE	Image Capture	$600.00		$0.00	1	0	$0.00	$0.00	
Photo Identification System Printer/Laminator	Fargo	Pro-L	$7,440.00		$0.00	1	0	$0.00	$0.00	
										$0.00
Hourly Labor Rate			$65.00							
Total Parts and Labor									$0.00	$0.00
Estimated Sales Tax									$0.00	$0.00
Estimated Freight									$0.00	$0.00
Estimated Permits									$250.00	$250.00
TOTAL ESTIMATED COST									$250.00	$250.00

each subsystem and the total systems, often in a simple narrative format. Later budgetary estimates are more detailed, like the one in Figure 4.

4.3.6 DESIGN TEAM

Security design does not exist in a vacuum. Security managers and directors need to determine who in their organizations should be involved in the design process and what their relationship should be with the design and construction professionals. The players should be identified early in the planning process—as soon as the extent of the project is known—so they can contribute to the initial preliminary design process and benefit from knowledge of it. The management style of the organization will determine the selection of those team members, as will the site and nature of the project. Not all of those in the following list will necessarily be included in the team. For example, the design team roles of the CEO and CFO may be performed by their delegates. The following are possible members of the design team:

- **Chief executive officer.** The CEO is involved in the project for two major reasons. The first is to ensure that the goals of the security program reflect the corporate mission and that the corporate image is maintained or enhanced. The second is to provide top-down support for the security program. Nothing can scuttle a well-designed program more quickly than lack of interest by executive management.

- **Chief financial officer.** The CFO keeps an eye on the cost/benefit factor of the project and approves funding once it is justified. The CFO is accustomed to reviewing quantitative data and bases investment decisions on returns.

- **Human resources manager.** Some security procedures are managed by the human resources department, such as careful hiring and firing practices and maintenance of an access control system cardholder database. Also, in many organizations the security function reports to the human resources manager.

- **Information technology manager.** Information technology is both an asset and a vulnerability. Particularly where security systems operate on an enterprise network, the information technology department should be involved to ensure that corporate standards are maintained.

- **Facilities manager.** In larger organizations, facilities management is closely aligned to the security function, especially regarding security technology and systems monitoring.

- **Project architect.** For projects involving major construction, an architect is usually involved.

- **Construction manager.** Larger construction projects may include a specialist firm that is responsible for all construction and implementation. The construction manager usually gets involved early in the design process to keep an eye open for constructability, design approval, and cost issues.

- **Security system designer.** Security subsystems and their integration are becoming increasingly complex, so considerable experience is required to design a system that addresses both vulnerabilities and operational issues. In larger organizations this capability may be available in-house. However, since new system projects do not occur often within an organization, it may be useful to retain a consultant who has relevant experience and no vested interest in any service or equipment being provided.

- **Security manager.** The security manager's role is essential to the successful design and implementation of the system. The security manager should understand these important concepts:

    The system does not belong to the architect, security consultant, system vendor, users, or even CEO or shareholders. The security manager lives with the consequences of system failure and therefore must assume accountability and ownership of the system. Ownership is achieved by understanding the process of design and implementation and by maintaining direct involvement throughout the project.

    One person cannot master all aspects of design and construction. Security managers who resist hiring specialists may end up relying on the inadequate expertise of company employees or security vendors. The security manager should identify where expertise is lacking and be prepared to hire specialists—as employees or consultants—to to help him or her maintain the level of involvement required to achieve ownership.

    To be successful, the system solutions must reflect the organization's mission, must be responsive to the organization's culture and business operations, and must have executive management's approval and involvement. For these reasons, the project team should prepare a security business case using terms and approaches common to other corporate capital investment projects.

4.4. DESIGN AND DOCUMENTATION PHASE

Next the project moves into the design and documentation phase. In the construction design industry, this may be split into two phases, the design development phase and construction documents phase. Alternatively, it may be considered as a single phase called construction documents, or CD, with the completion of the design development work being referred to as 30, 35, or 50 percent CD. Generally, design development includes a preliminary design (30 to 35 percent) following the conceptual or schematic design and concludes with a 50 to 60 percent design development. The percentages represent the level of completion of the final construction documents.

Following design development, CDs usually begin with a 60 percent design and pass through a 90 percent CD phase submission to conclude with a 100 percent CD set. If the security designer is working with an architect, design phases and submissions usually coincide with those reflected in the contract between the architect and the owner. However, if the security designer is working without an architect, these design submissions are usually tailored to the specific project and almost always include a conceptual, preliminary, and construction set submission with corresponding design reviews at each phase.

The objective of the design and documentation phase is to complete the design and to document the process to the level of detail necessary for the chosen method of procurement. A greater level of detail in the design will lead to better responses from bidders and lower project costs.

The complete set of procurement documents, known as contract (or bid) documents, will consist of three sections: contractual details, construction specifications, and construction drawings. In a procurement of services (such as guard services), the third section is not required. In a construction-related procurement (involving, for example, access control and CCTV), the specifications and drawings are called construction documents. On smaller projects, it is common to see all the written specifications included on the construction drawings.

4.4.1 CONTRACTUAL DETAILS

This section of the contract documents describes the form of contract to be used when a supplier has been chosen. It covers insurance and bonding requirements, site regulations, labor rules (union or non-union, wage rates, etc.), delivery and payment terms, methods of measuring work progress for partial payment, owner recourse in the event of nonperformance, termination conditions, application of unit pricing to additions and deletions, instructions to bidders, etc. For a large construction project, the architect or the owner's construction manager develops this document to cover all trades, including security. For smaller jobs, the company's purchasing department may develop this section. In most cases, the document is included in the contract documents and is modified to suit the particular project as the project progresses.

4.4.2 SPECIFICATIONS

The security systems mirrors and complements the actual systems design in such detail that:

- The final implementation reflects what was intended. In all cases, the systems specification contains the actual performance instructions and criteria for constructing the systems included in the design. Included in the specification should be functional testing to ensure the system will do what it is designed to do as well as a continual periodic programmed testing to ensure the integrity of the system over time. Drawings and plans are virtually useless and are open to interpretation unless there are associated specifications detailing construction and systems performance criteria. Drawings and plans show what is to be constructed, whereas the specification details the owner's intent and how it is to be constructed.

- All bidders get the same, complete understanding of the requirements. Incomplete or inaccurate specifications can lead to wildly different bids and an inability of the procurer to compare them.

Because of the level of detail required, specifications tend to be wordy and very technical. Considerable technical experience in the design, procurement, construction, and operation of a security system is needed to prepare good specifications. With poor specifications, vendors may make quality and performance choices for the owner without the owner's knowledge until the system is installed and operating.

Boilerplate specifications are available as a starting point for customization to meet project-specific requirements. Most experienced security system designers have developed their own master specifications. The specification should reflect lessons learned from previous security system projects. For example, a contractor may have misinterpreted a phrase in the specifications, leading to reduced functionality of the system or increased costs.

Specification sections are numbered depending on the construction trade so that each section can be issued separately. Standard specifications are available from the American Institute of Architects (www.aia.org) and the Construction Specifications Institute (www.csinet.org). For example, the Construction Specifications Institute publishes MasterFormat™ and MasterSpec™ standards. A project manager or architect would use the various divisions of those standards to document an entire construction project. Electronic Safety and Security is Division 28 of the Facility Services Subgroup.

Especially with the trend toward integration among subsystems and procurement of all security systems through a single contractor, it is common to depart from this format and include all the security systems within a single, custom-designed section. Most architects and project managers prefer the security systems all in one specification.

Each individual specification section consists of a standard format divided into three parts: **general, products, and execution**. Each part is divided into subsections and sub-subsections. Not all titles are applicable to every project, so the specification format is often modified by the security designer to suit the unique circumstances of the project. The importance of the standard format is to ensure the following:

- The final specifications are complete in all details.

- Contractors can easily find specific details when preparing a proposal or bid or when implementing the system.

A security system specification should include the following, at least:

- **instructions to bidders** with a list of all documents included in the contract documents
- **list of project references**
- **functional description** of the complete systems design, its intended functional operation in a concept of operations, maintenance and warranty requirements, quality assurance provisions, and installation schedule
- **list of design drawings**
- **list and description of products and services to be included in the contract**
- **list of required products and services included in other contracts** (such as electrical door hardware, which is provided and installed under the door hardware contract but must be connected to the security system by the security contractor)
- **list of applicable codes and standards**
- **support services**, such as drawing, sample and documentation submittals, commissioning, testing, training, warranty, maintenance, and spare parts
- **technical descriptions** of all major subsystems and their components, including capacity, capability, expandability, performance and operational parameters, environmental operating parameters, installation and integration details, appearance and finish, and acceptable makes and models
- **general site conditions**, installation standards and quality control standards

4.4.3 DRAWINGS

Along with specifications, drawings are the cornerstone of any construction project. A picture or diagram of design intent is less likely to be misinterpreted by contractors. However, to avoid ambiguity and to manage any discrepancies among the drawings, specifications have precedence over drawings.

Most drawings are produced by computer-aided design drafting (CADD) systems. Compared to manual drawings, CADD files are clearer, modifications are quicker and less expensive to make, and documents can be shared more easily. In addition, many security designers themselves work directly with CADD systems rather than making sketches for a draftsperson to convert into a finished drawing. The direct approach eliminates transcription errors and the need to train an additional person on project engineering requirements.

Security system drawings usually consist of plans, elevations, details, risers, and hardware schedules. Each drawing is either a site plan or floor plan showing the security systems devices by type and location. The floor plan in Figure 5 is one such drawing.

Plans

Each security systems plan shows a top-down, map-like view of an area where security devices and systems are located. The area may be a complete site, a building floor, or part of a floor. Many plan drawing sheets as needed to show all areas where security systems will be installed. The background information on a plan drawing consists of such items as fence lines, building wall locations, interior partitions, doors, furniture, door and room numbers (known as targets or tags), room names, floor materials, stairs, fixed equipment, etc. The architect usually provides the background drawings for a construction project. For a system upgrade project, the company may already have background drawings. For manual drafting,

Figure 5
Typical Floor Plan Drawing

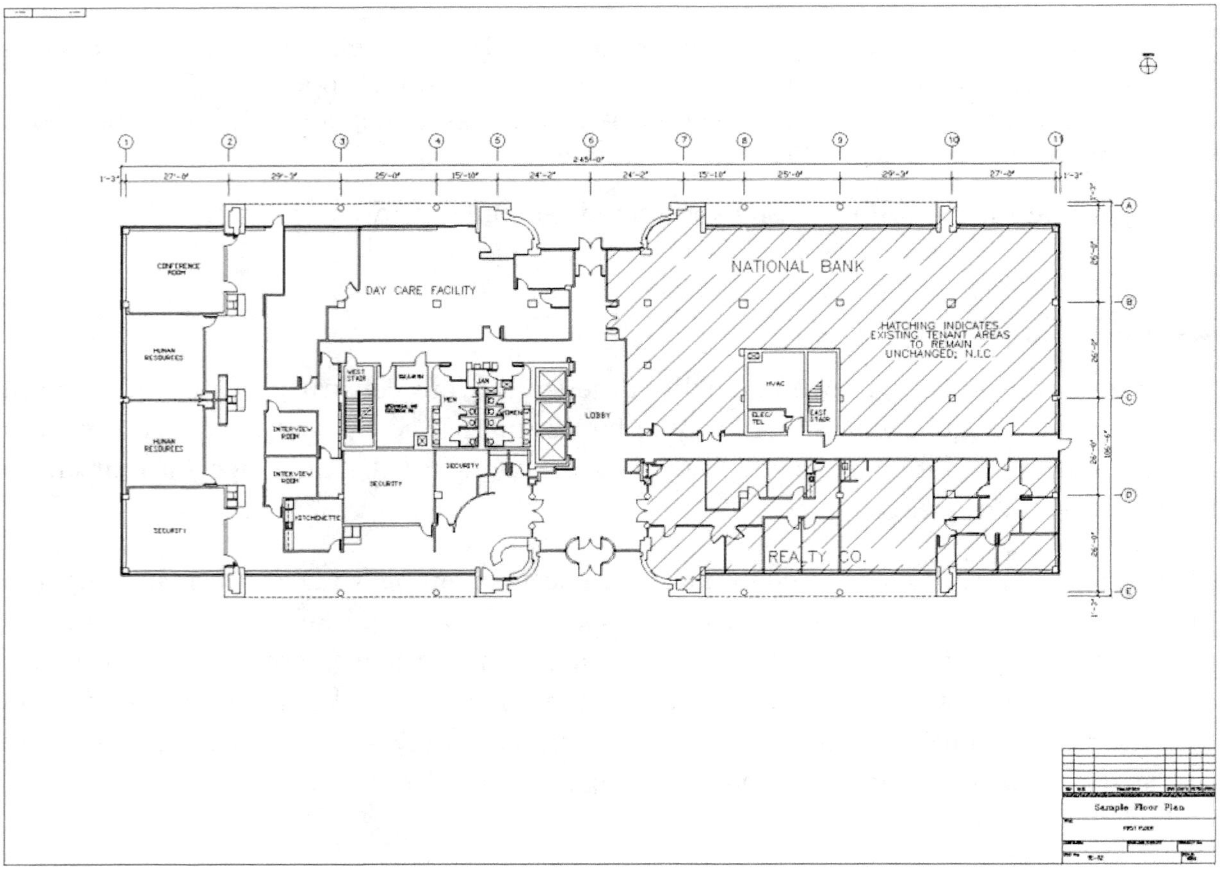

the background information is provided on transparent (also known as reproducible) sheets, such as paper vellum or Mylar, onto which are drawn the symbols that represent the various items of security equipment and, in some cases, lines between equipment to show interconnections. The level of background detail must be sufficient but not so extensive that the drawings become busy and security equipment becomes inconspicuous.

For CADD drafting, the background drawing file consists of a number of layers (for example, one each for walls, doors, furniture, and lighting design). The CADD draftsperson can select which levels are required and turn them on or off. Security symbols are usually kept on their own layer and are copied to required locations as predefined blocks. If the architect changes the background design, the old security layer can be superimposed on the new architectural background. Changes to the security layer only need to be made when security is affected by the architectural change, such as new or relocated alarm doors.

Many individual companies, security magazines, architects, engineers, security consultants, and standards-making organizations have developed sets of security symbols. The most common symbols set for manual drafting is issued by the American Society for Testing and Materials in *Standard Practice for Security Engineering Symbols, F967-03* (2007). In 1995 a new standard for symbols was developed jointly by the International Association of Professional Security Consultants and the Security Industry Association. Titled *SIA Architectural Graphic Standard—CAD Symbols for Security System Layout, SIA AG-01-1995.12*, the standard provided symbols that were incorporated into the ASTM standard. Whichever set of security symbols is used, the specifications should require that the same set be used for contractor-submitted drawings. Figure 6 presents a drawing detail showing symbology to depict security devices. It also depicts a numbering scheme for security devices for later reference in schedules.

Elevations

Elevations are views of vertical surfaces and are included to show mounting heights and locations of wall-mounted devices, such as cameras, card readers, and motion sensors. Elevation backgrounds can be provided by the architect or from the organization's files. A sample security door elevation is shown in Figure 7.

Details

Most plans and elevations are shown in small scale for the drawings (for example, 1/8 inch equaling 1 foot). Detailed drawing sheets can be developed to define elements of the system in more detail. Such details may include special mounting techniques, custom part design dimensions, or cable terminations. These are usually developed specifically for a project. However, a security system designer may have access to drawings from previous projects that can be reused or modified.

Figure 6
Typical Drawing Device Symbology

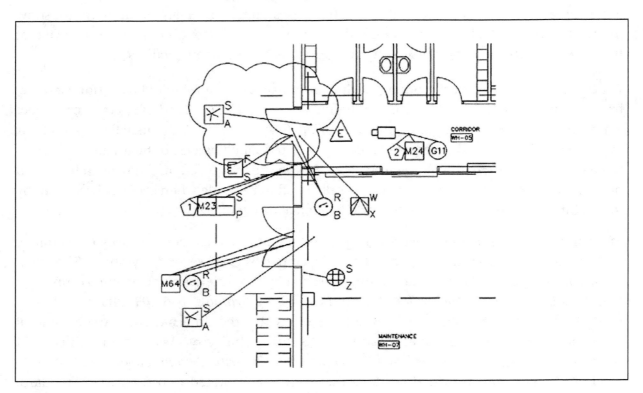

Risers

Riser diagrams are representations of complete subsystems, such as CCTV or access control. They schematically demonstrate all the associated devices and components and their interconnecting cables. For a campus environment, each building may be shown as a different block. For a high-rise building, each floor may be shown in a vertical, elevation-like format. On smaller projects, all subsystem riser diagrams, with their interconnections and interfaces, may be placed on a single sheet. Because so much information is depicted on a single drawing, it is used by designers and contractors as the master drawing. In particular, contractors tend to use the riser diagrams for device counts when developing their bid price for the project. For these reasons it is very important that riser diagrams be accurate and complete. Figure 8 shows a sample of a small riser diagram.

Hardware Schedules

Hardware schedules are tables of related security devices. They provide detailed information that cannot easily be shown on drawings or in the text of a specification. Schedules are used for door hardware, control devices, intrusion sensors, cameras, monitors, and other devices that appear repetitively, such as termination panels. Figures 9 and 10 show sample hardware

INTEGRATED SECURITY SYSTEMS DESIGN AND SPECIFICATION

Figure 7
Security Door Elevation

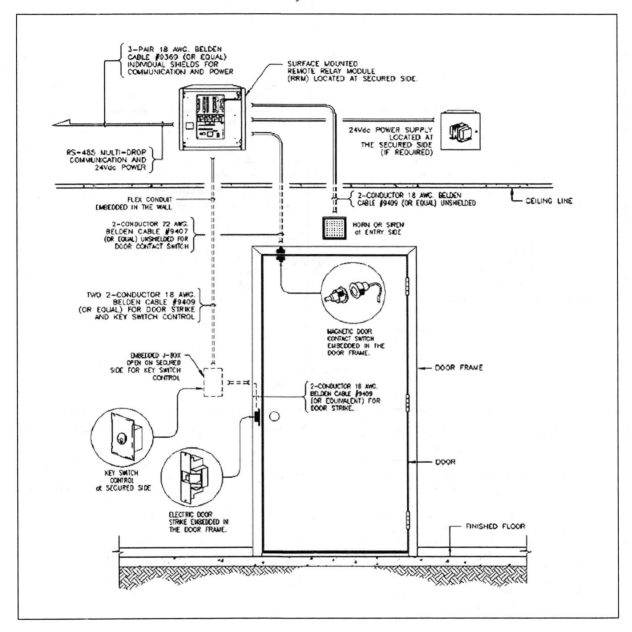

schedules for security door-related devices and CCTV cameras. The schedules are often shown on security drawings but may also be appended to security system specifications.

4.4.4 DESIGN COORDINATION

Security design on a construction project is affected by many other design disciplines. Careful coordination among the security system designer and other design team members is

Figure 8
Sample Riser Diagram

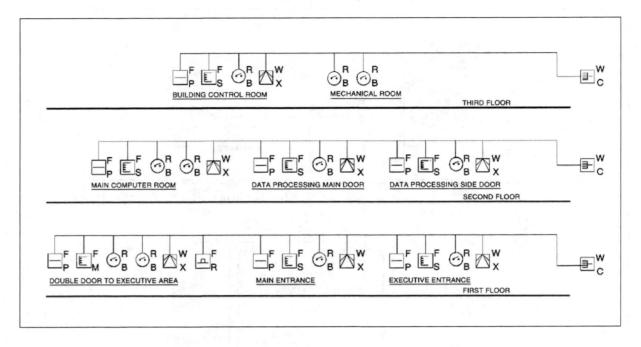

Figure 9
Sample Door Schedule

#	FLOOR	LOCATION/ROOM	DOOR	ACCESS	ELECTRIC LOCK	OTHER HARDWARE
B01	Basement	Loading Dock				Intercom
B02	Basement	Shipping/Receiving/Storage	DBL	Card reader	Strike	2 door contact (recessed), request to exit
B03	Basement	Shipping/Receiving office	SGL	Card reader	Strike	2 door contact (recessed), request to exit
B04	Basement	Shipping/Receiving office				Intercom
B05	Basement	Shipping/Receiving office				Monitor
B06	Basement	Shipping/Receiving office				Telephone
B07	Basement	Shipping/Receiving office				Duress

essential to avoid missing elements of the design or procuring things twice. Some elements of the security system will be procured and described in specification sections that are prepared by other design disciplines. For example, in new construction it is common for all electrical power, including that required for the security system, to be specified in a single specification and procured and installed by a single electrical contractor. Listed below are the various design team members with whom the security designer is usually required to coordinate.

INTEGRATED SECURITY SYSTEMS DESIGN AND SPECIFICATION

Figure 10
Sample Camera Schedule

CAMERA NUMBER	LOCATION / VIEW	TYPE	LENS	HOUSING	MOUNT	HEIGHT	ALARM CALL-UP
B1	Loading dock and area leading into the shipping/receiving area	PTZ	10:01	Enviro DOME	Wall	12 feet	
C01	Ground floor lobby. The field of view of this camera includes the lobby and the corridor leading to the east exit.	PTZ	3.2 mm - 6.4 mm varifocal	DOME	Ceiling	Ceiling	Yes
C02	Third floor lobby. The field of view of this camera includes the lobby.	Fiixed dome (Sensormatic AD614LSP or equivalent)	3.2 mm - 6.4 mm varifocal	DOME	Ceiling	Ceiling	Yes

Architect

The architect lays out the space within a facility. Any space required by the security system, such as a console, locker rooms, riser closets, or security equipment storage rooms, must be coordinated with the architect. The earlier this occurs in the design process, the more likely the security department will get the space it needs.

The architect usually specifies door hardware. In addition, the architect ensures that appearances and finishes are consistent and that any door cut-outs required for hardware installation are performed in the factory.

Electrical Engineer

The primary coordination concern with the electrical engineer is ensuring that main electrical power is provided at all locations where security equipment requires it. Dedicated circuit amperage and electrical support requirements (such as a generator or uninterruptible power supply) need to be specified.

Similarly it is common to include electrical back boxes, junction boxes, and conduit in the electrical section of the specifications. These are installed by the electrical contractor.

If the electrical engineer is designing a separate fire alarm system, its interface to cut power to fail-safe security door locks, as required by code, needs to be fully coordinated.

Mechanical Engineer

This coordination issue relates to heating, ventilating, and air conditioning (HVAC) requirements for security spaces. The mechanical engineer needs data on heat loads and duration of occupancy (such as a 24-hour security control room or security equipment with special environmental needs) to ensure that the required environment is provided. If conduit for security cabling is required above the finished ceiling level (and its runs are not being designed by the electrical engineer), locations need to be coordinated with HVAC ductwork.

Vertical Transportation Designer

Security equipment associated with elevators, either inside or outside the cabs, requires careful coordination. The placement and mounting of CCTV cameras in cabs is critical to their effectiveness and very dependent on cab design. The designer may be on the staff of the architect or mechanical or electrical engineer or may be a specialist consultant. Other coordination issues are the inclusions in traveling cable of the security equipment's needs, power requirements on the roof of the cab, and any interfaces required in the elevator machine rooms. Use of security equipment with escalators must also be coordinated.

4.4.5 CONSTRUCTION DOCUMENT REVIEW, APPROVALS, AND ISSUE

During the development of the construction documents, it is common to set certain milestones at which progress is reviewed. The milestones may be target dates or nominal percentages of completion, such as 35, 60, 75, 95, and 100 percent. Smaller projects should not require as many reviews. At each milestone, it is important to compare the state of the design with the original security requirements and design criteria to ensure that the vulnerabilities originally identified are being addressed by the security system and that the security program objectives are being met. As the design progresses, the construction cost estimate should be updated to confirm that the project remains on budget. The estimate should become more accurate as the design gets closer to finalization.

When affected parties are brought into the review, their scope should be limited to the portion of the design that affects them. Some project team members may not be familiar with construction documents and may find them difficult to understand. It is often beneficial for the security designer to provide a presentation of the security system design concepts and intended operations. Comments and requests for changes should be provided in writing, and the response, together with details of any schedule or cost implications, should also be documented. Changes made early in the design process have far less impact on cost and schedule than do those made later. Also, changes made during the design phase are less expensive to implement than those made during or after installation. Changes are inevitable, but major or frequent changes may indicate a project with an incomplete or inaccurate design.

If formal approvals are required, either at each milestone or only at completion, they should be obtained before issuing final construction documents. For major construction projects, the final documents may need to be stamped—that is, imprinted with the seal of a professional engineer (PE) or architect. For security systems design stamping, it is usually the electrical PE designer who is called on for a stamp. If the security consultant/engineer does not have a PE license, he or she should be an employee of, or work very closely with, a firm that has employees with PE licenses. No reputable professional engineer will rubber stamp construction documents because design liability passes to the professional who stamps the drawings. The professional engineer needs to have been involved in the design process and must perform an extensive design review (and such services are not inexpensive). Since security system design work typically relies on low-voltage electrical systems, the need to stamp security drawings is infrequent and the added expense is usually unnecessary.

The completed set of contract documents (contractual details and construction documents) may be issued to bidders by the owner organization, the security consultant, or the owner's architect or construction manager. Contractual details and specifications and equipment schedules are usually produced on letter-size paper and can be photocopied. Final drawings are usually large sheets from which blueprints are made. It takes time for reproductions to be made and issued. Some government organizations that publicly advertise projects require bidders to collect and pay for sets of construction documents.

4.4.6 PROCUREMENT PHASE

The three major forms of security systems procurement are sole source, request for proposal (RFP), and invitation for bid (IFB), with some variations depending on whether the buyer is a government agency or commercial firm. Each form of procurement has its benefits, but the type should be selected before or at the start of the design phase. The reason is that the type of procurement affects the level of detail required in the construction documents. If an owner already has a vendor on board, a sole source procurement is appropriate and the level of detail of the design should complement the knowledge already held by the vendor. If, however, a vendor is to be chosen competitively on a wide variety of factors, such as cost, schedule, technical ability, etc., then a request for proposal is the appropriate procurement form. The vendor will require project details sufficient to submit a responsive proposal, and the owner will require sufficient vendor details to make an appropriate selection. Invitations for bid typically require sufficiently detailed design information for the responding vendors to offer a firm fixed price to install and commission the systems specified. Since IFBs key on a vendor's price, the owner must make absolutely sure that sufficient design details and instructions are provided so as not to leave any loopholes allowing vendors to substitute inferior or inadequate equipment merely to win the job.

Some large organizations have the capability to install, commission, test, and maintain their own security systems. Although their design phase may be extensive and detailed, their procurement phase may be as simple as issuing purchase orders for hardware at pre-negotiated prices to pre-qualified vendors.

4.4.7 SOLE SOURCE PROCUREMENT

For small projects, this may be the most appropriate method of procurement. The organization pre-qualifies a reputable security system contractor, works with the contractor to design the system, and negotiates the cost of equipment, installation, and service. On the positive side, the construction documents are usually simple, reducing owner design costs and saving time. On the negative side, there is a tendency to focus on hardware and technology only, leaving the equally important personnel, procedures, and facilities subsystems for others. Also, the owner may tend to skip the all-important security planning process and rely on advice from a contractor with a vested interest in selling equipment. In addition, without a competitive bidding process, the organization has no means of comparing prices. This method of procurement is recommended only where the security owner has the capability to perform the security needs analysis and has good prior knowledge of systems and prices.

4.4.8 REQUEST FOR PROPOSAL

The RFP is almost always based on a set of detailed design and construction documents. The specifications are usually generic and performance-based. Equipment makes and models are often listed with the phrase "or approved equal." In some cases, specific models may be mandated for compatibility or commonality with existing equipment. Overall, in the RFP process the owner typically procures a security business partner, not just a one-time security systems installer.

An RFP response may be open to any contractor or it may be limited to a list of prequalified contractors. In addition to providing a cost proposal, a proposer must submit a technical proposal that describes the firm's understanding of the requirements and how the objectives will be met. It is common to allow responders to propose alternative solutions, called "alternates." To sensibly compare cost proposals from different contractors, it is usually necessary to require the contractors to respond to the specified design and then, if they wish, allow them to provide alternates as additional solutions. It is not uncommon to instruct proposers that alternates must produce some definable improvement in performance and be of equal or lesser cost than the base bid. The owner then benefits from the experience of the contractor while maintaining full control over the design process. The organization may select one or more of the proposers to participate in final negotiations.

The RFP need not restrict the organization to accept the lowest bid. Instead, it aims to obtain the best value. Value may be defined by the organization to suit its needs, but it should include such factors as price, quality, experience, and schedule. If price will not be the determining factor in vendor selection, the RFP should say so.

INTEGRATED SECURITY SYSTEMS DESIGN AND SPECIFICATION

A contractor's response to an RFP usually takes longer to prepare than responses to other types of procurement because both a technical and a cost proposal must be prepared. Three to four weeks is the typical minimum proposal preparation time for medium-size to large projects.

4.4.9 INVITATION FOR BID

IFB is commonly used by government and other organizations whose procurement procedures require that projects be competitively bid and that the award be given to the lowest qualified, responsive bidder. No technical proposals or alternative solutions are sought, so the construction documents must be extremely explicit. The onus of selecting equipment makes and models, and accuracy of the security system design, is placed solely on the design team. Bidders submit a cost proposal or bid, which may contain unit pricing and whatever price breakdown is requested. Bidders may also need to show their qualifications. The award is then made, usually without negotiation, to the lowest qualified bidder who has conformed to the bidding instructions.

The IFB requires additional time and cost in design and specification, but typically needs only one to two weeks of procurement time, depending on the size and complexity of the project. It is common to require bids to be sealed and delivered by a specific time to a specific location. At the time and place, the bids are opened (often publicly) and the apparent winner is announced. Contracts are signed when the apparent winner's proposal has been checked for completeness, accuracy, and qualifications.

4.4.10 PROCUREMENT PROCESS

It may be important to hold a pre-bid conference to which a representative of each contractor is invited. At that conference, the owner or the owner's consulting engineer provides a complete review of the bid documents and a walk-through of affected buildings and locations. If applicable, the conference can be held at the site where the new security system will be installed so that bidders can see the field conditions. The conference should be held approximately one week after the construction documents have been issued for bid. This gives the bidders enough time to review the documents but allows time to incorporate additional information into the proposal if necessary. All questions and answers at the conference should be recorded by a design team representative in the meeting minutes. Any questions from contractors after the conference should be asked in writing, and the answers should be transmitted to all prospective contractors. It is best to set a deadline of a week before the proposal due date, after which questions are no longer taken. A single point of contact should be nominated for all questions.

Once contractor proposals or bids have been received, they need to be checked for completeness and accuracy. The contract details may say that any inaccuracies or incompleteness in the proposal will cause it to be rejected, but most commercial organizations do not reject bids unless they show signs of incompetence or gross incompleteness. It is useful to develop a matrix with the column headings representing the contractors and the row headings listing the main security system features and components. The matrix helps the reviewer check that the technical proposals of each responder have addressed all aspects of the construction documents. A similar matrix can be developed to compare price proposals and any alternates.

When comparing proposal costs, the life cycle cost of each proposed system should be calculated. The first step is to identify the specific objectives and goals you want the system to perform, and how long the life expectancy of the system. In its simplest form, the life cycle cost is the sum of the capital cost and the maintenance cost over the useful life of the system. Typically, maintenance and warranty costs equal 11 percent of the total capital systems construction cost. Calculating those figures can reveal whether the low bidder has priced the system at a low profit margin but plans to make up the difference in high charges for maintenance.

If one proposal's system costs are much lower than those of the other proposals, the low proposal should be scrutinized carefully for the following:

- mathematical errors
- quality of equipment being proposed
- experience of the contractor on projects of this size and complexity
- contractor's understanding of the project
- financial stability of the contractor

All contractors' references should be checked before an award decision is made.

Interviews with the leading contenders may be revealing. In particular, the designer should request that each contractor's project manager and site supervisor (possibly the same person) be present at the meeting. The designer should attempt to determine the following:

- Is there good chemistry with the contractor's representatives?
- Do they have the experience and power of personality to work well with the other trades on the project?
- How have they resolved problems that occurred on other projects?

It also helps to find out what other clients think about how problems were resolved.

Negotiating the final price with the short list of contractors, if permitted by the procurement regulations, should be done on the basis of value. If the contractor's profit margin is too small, quality and responsiveness will suffer. A good contractor with a realistic profit will go the extra mile to ensure that implementation problems are solved and that all parties will be able to look at the finished implementation with pride. In the end, it is beneficial for the owner and contractor to enter into a business partnership, not a one-time sale.

REFERENCES

American Society for Testing and Materials. (2007). *Standard practice for security engineering symbols*. F967-03. West Conshohocken, PA: ASTM.

Security Industry Association. (2000). *SIA architectural graphic standard—CAD symbols for security system layout, SIA AG-01-1995.12*. Alexandria, VA: